DISCLAIMER

This book is a reprint of Austin M. 2nd ed.: Questions and Answers on Rats, Mice and Other Vertebrate Pests. Duluth, MN: Advanstar Communications, 1999.

This book contains historical information and is an exact reprint except for the following: 1. Color images have been converted to black and white, 2. Disclaimer, and 3. Information about the present publisher (Wildlife Control Consultant, LLC and advertising information about the publisher.

Readers should understand that this information is historical. Laws, practices, and scientific information change. Readers are admonished that the book is sold with no warranty of present accuracy or alignment with present scientific understanding or legal standards.

Readers are responsible to follow all federal, state, and local laws. This book does not contain all information available on the subject and has not been created to be specific to any individual's or organizations' situation or needs. Every effort has been made to make this book as accurate as possible. However, there may be typographical and or content errors. Therefore, this book should serve only as a general guide and not as the ultimate source of subject information. Readers acknowledge that working with wildlife is dangerous and can result in severe injury and even death. The authors and publisher shall have no liability or responsibility to any person or entity regarding any loss or dam- age incurred, or alleged to have incurred, directly or indirectly, by the information contained in this book. You hereby agree to be bound by this disclaimer, or you may return this book within 30 days of the date of purchase for a full refund minus shipping, damage deductions, and transaction costs.

 Wildlife Control Consultant, LLC
WildlifeControlConsultant.com
©2019 Wildlife Control Consultant, LLC.

THIRD EDITION

Wildlife Control Consultant, LLC is revising the commensal rodent control portion of this book. Buyers of this edition are eligible for a 20% discount and free shipping to the continental U.S. To be eligible, simply e-mail a copy of the purchase receipt along with mailing address to wildlifecontrolcontrolconsultant@gmail.com and we will put you on the list. Purchases must be made within 30 days of notification of the book's availability.

Vertebrate Pest Handbook

Second Edition

Austin M. Frishman, Ph.D.

an ADVANSTAR publication

DEDICATION

To Al Hochman of Clover Exterminators—The man who taught me the importance of practical pest control and first made me aware of the need for professional training in the pest control industry.

Copyright © 2019 Wildlife Control Consultant, LLC

Copyright © 1999 Advanstar Communications

All rights reserved. No portion of this book may be reproduced or used in any form or by any means—graphic, electronic, or mechanical, including photocopying, recording, taping, or information storage and retrieval systems—without written permission of the publisher.

Printed in the United States of America

10 9 8 7 6 5 4 3 2 1

ISBN 0-929870-51-4 Old New ISBN 978-0-359-77318-3

Published by Advanstar Communications, Inc.

Advanstar Communications is a US Business information company that publishes magazines and journals, produces expositions and conferences, and provides a wide range of marketing services.

For additional information on any magazines or a complete catalog of Advanstar Communications books, please write to Advanstar Communications Customer Service, 131 West 1st Street, Duluth, MN 55802 USA. To purchase copies of the *Vertebrate Pest Handbook,* please call 218-723-9180. Price per copy is $31.95; quantity discounts available.

Interior Design: Lachina Publishing Services, Inc., Cleveland, OH
Product Manager: Heather Gooch and Danell Durica

Other Publications
by
Wildlife Control Consultant, LLC
Dealer Inquiries Welcomed

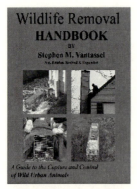

Wildlife Removal Handbook, 3rd ed.

205 pages of information on how to start and run your wildlife control operator business. Helpful for those who just want to learn how to control raccoons, skunks, squirrels, woodchucks and more like a professional.

The Wildlife Damage Inspection Handbook, 3rd ed.

180 letter-sized pages. Color

This text teaches the how to read and interpret vertebrate animal sign and damage to structures and lawns and gardens. Don't guess what caused the problem, know what species caused the problem.

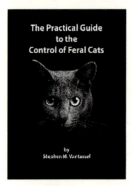

The Practical Guide to the Control of Feral Cats

103 pages of effective control information. Free-ranging cats are a growing environmental and public health threat around the country. Learn how to do your part to resolve human-feline conflicts.

Wildlife Control Consultant, LLC
WildlifeControlConsultant.com
©2019 Wildlife Control Consultant, LLC.

Contents

Preface	vii
Acknowledgments	vii
Introduction	ix

CHAPTER 1   Rodents in General   1

CHAPTER 2   Norway Rats (*Rattus norvegicus*)   8

CHAPTER 3   Roof Rats (*Rattus rattus*)   22

**CHAPTER 4   House Mouse
(*Mus musculus* or *Mus domesticus*)**   28

CHAPTER 5   Rodent Diseases

  Human Diseases and Parasites Associated with
  Rodents   38
    General Material

  Diseases Transmitted by the Rodents Themselves
    Hantavirus   39
    Leptospirosis   46
    Lymphocytic Choriomeningitis   47
    Rat-bite Fever   48
    Salmonellosis   49
    Tapeworms   50
    Trichinosis   51

  Diseases Transmitted by Arthropod Parasites
  of Rodents
    Lyme Disease   51
    Human Granulocytic Ehrlichiosis (HGE)   67
    Murine Typhus Fever   68
    Plague   69
    Rickettsialpox   70

CHAPTER 6 Non-Commensal Rodents

Cotton Rats (*Sigmodon*)	72
Deer Mice and White-Footed Mice (*Peromyscus*)	74
Meadow Voles (*Microtus*) and Pine Voles (*Pitymys pinetorum*)	77
Muskrats (*Ondatra zibethica*)	81
Nutria (*Myocastor coypus*)	83
Pocket Gophers (*Geomys* and *Thomomys*)	85
Porcupines (*Erethizon*)	88
Prairie Dogs (*Cynomys*)	90
Squirrels and Chipmunks (*Sciuridae*)	91
Chipmunks (*Tamias*)	97
Woodchucks (*Marmota monax*)	100

CHAPTER 7 Pest Vertebrates Other Than Rodents

Armadillos (*Dasypodae*)	103
Bats (*Chiroptera*)	106
Moles (*Talpidae*)	118
Opossums (*Didelphis virginiana*)	124
Rabbits (*Sylvilagus*)	126
Raccoons (*Procyon*)	130
Shrews (*Soricidae*)	138
Skunks (*Mustelidae*)	141
Snakes (*Colubridae*)	146

CHAPTER 8 Non-Toxic Rodent Control Methods

General Considerations	151
Rodent Stoppage	156
Rodent Odors	159
Traps	160
Multiple-catch Traps	165
Glue Boards	167

Repellents	172
Sound Devices for Rodent Control	173
Black Lights for Rodent Contamination Detection	173

CHAPTER 9—Rodenticides

General Considerations	175
Tracking Powder	188
Zinc Phosphide	193
Bromethalin	196
Cholecalciferol	197
Anticoagulants—General Considerations	198
Baiting at Dumps	204

CHAPTER 10—References for Training

Videos	206
Cassettes	207
Slides	207
Published Material	207
Training Courses	210
Index	211

OTHER PUBLICATIONS BY WILDLIFE CONTROL CONSUTLANT, LLC

Dealer Inquiries Welcomed

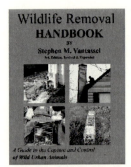

Wildlife Removal Handbook, 3rd ed.

205 pages of information on how to start and run your wildlife control operator business. Helpful for those who just want to learn how to control raccoons, skunks, squirrels, woodchucks and more like a professional.

The Wildlife Damage Inspection Handbook, 3rd ed.

180 letter-sized pages. Color

This text teaches the how to read and interpret vertebrate animal sign and damage to structures and lawns and gardens. Don't guess what caused the problem, know what species caused the problem.

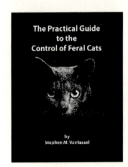

The Practical Guide to the Control of Feral Cats

103 pages of effective control information. Free-ranging cats are a growing environmental and public health threat around the country. Learn how to do your part to resolve human-feline conflicts.

Wildlife Control Consultant, LLC
WildlifeControlConsultant.com
©2019 Wildlife Control Consultant, LLC.

The *Vertebrate Pest Handbook* by Dr. Austin Frishman is funded by Zeneca Professional Pest Control Products.

ZENECA RODENTICIDES

Weatherblok® XT rodenticide is a new addition to the Talon® Rodenticide family. Containing brodifacoum, the most active anticoagulant, Weatherblok has a new extruded shape, new ingredients, larger mounting hole, and improved rodent acceptance.

As with all Talon products, Weatherblok contains bitrex, a human taste deterrant, and has a blue warning color. Weatherblok and the other Talon products have the broadest label of any second-generation materials, including effectiveness against both resistant rats and resistant mice, and outdoor use in non-urban areas and in burrows.

As with all anticoagulants, vitamin K is an antidote in the event of accidental ingestion.

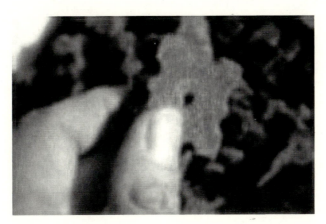

See photo section for color version of this photo.

Preface

The author welcomes comments and solicits questions for future publications. No one knows more about questions that need answering than the people working in the field.

It is difficult to imagine that 21 years have passed since the first edition of this book. Some things never change, and so remain in this book. The majority of items are new and reflect the changes in attitude and technology.

It is recognized that deer, wild cats and birds are also rapidly increasing their importance as a wildlife nuisance problem. The author chooses to focus on a more limited, but detailed vertebrate arena.

Acknowledgments

Special thanks go to Dr. Robert Corrigan and Mr. Dale Kaukeinen, who spent many hours reviewing the manuscript. Both, through their knowledge and personal concern, have significantly elevated the contents and flow of the book.

Introduction

Vertebrates are not insects, and therefore should not be treated as such when it comes to control. Yet many entomologists are charged with controlling these four-legged vermin along with the six- and eight-legged variety. Training on vertebrate pest problems is often limited for pest control operators (PCOs) and restricted to hearsay and "the school of hard knocks."

The more we can learn about pest vertebrates, the better we can cope with them. Mammals of the same species exhibit individual variations in behavior, and may on occasion differ from the norm. Answers in this text are based on normal situations.

This book was written for several purposes.

- To serve as a handy reference for anyone engaged in vertebrate pest control.
- To aid both PCOs and people from the Board of Health in training new personnel.
- To assist professionals in answering the public on questions that arise on the job.
- To compile, under one cover, a working knowledge of most pest vertebrates.
- To provide test questions for supervisory pest control personnel who wish to train new and experienced employees.
- To stimulate all to think before they undertake a vertebrate pest control program.

The intent of the author is to strengthen the composition of this book with practical answers to practical questions. No one is more aware of the limits of this text than the author. It was impossible to cover every question that might arise. The topic of birds and their control has been completely omitted because to give it a cursory review could do more harm than good.

To aid the reader in finding the answer to a specific question, a detailed Contents and Index are provided. These references are divided into categories that should be understandable to those who do vertebrate pest control.

CHAPTER 1
Rodents in General

Define Each of the Following Terms

1. **Rodent**—Specific group of mammals characterized by chisel-shaped teeth. Rodent means "to gnaw."
2. **Commensal rodent**—A domestic rodent. One that lives in habitats occupied by humans. When you divide the word into *com* and *mensa,* it forms "together at the table."
3. **Feral animal**—A wild animal.
4. **Sylvatic rodent**—Any non-commensal rodent found in the forest or grassland.
5. **Altricial**—Born blind, hairless and helpless. Confined to the nest for some time after birth.
6. **Vole**—Includes several types of rodents including meadow mice, prairie mice and pine mice. The term "vole" once meant "mouse."
7. **Field mice**—Large group of mice, including pine mice and meadow mice (also called voles). Sometimes refers to house mice moving into structures in the late fall. There is no one species called "field mouse." In a field, you can have house mice, deer mice, white-footed mice and other rodents.
8. **Orchard mice**—Large group of rodents including pine mice and meadow mice (voles). Almost any mouse found in orchards.
9. **Deer mice**—Rodents belonging to the genus *Peromyscus,* along with white-footed mice, cotton mice, etc. Often a collective term for all *Peromyscus* mice.
10. **Pocket gophers**—Burrowing rodents belonging to the genera *Geomys* and *Thomomys.*
11. **Diastema**—Space between two teeth; the area behind a rodent's incisors.

Indicate with a check which of the following are rodents.

12. Prairie dog ✓
13. Skunk ✗
14. Cotton rat ✓
15. Short-tailed shrew ✗
16. Chipmunk ✓
17. Squirrel ✓
18. Pocket gopher ✓
19. Beaver ✓
20. Muskrat ✓
21. Porcupine ✓
22. House mouse ✓
23. Pine mouse ✓
24. Meadow vole ✓
25. Star-nosed mole ✗
26. Bat ✗
27. Opossum ✗
28. Norway rat ✓
29. Roof rat ✓
30. Florida water rat ✓
31. Cottontail rabbit ✗

List of numbers you should have checked above: 12, 14, 16, 17, 18, 19, 20, 21, 22, 23, 24, 28, 29 and 30.

Match Them Up!

Column A
Common Name
32. Roof rat _____
33. Polynesian rat _____
34. House mouse _____
35. Eastern chipmunk _____
36. Squirrel _____
37. Norway rat _____
38. White-footed mouse _____
39. Cotton rat _____
40. Deer mouse _____
41. Pocket gopher _____

Column B
Latin Name
a. *Peromyscus leucopus*
b. *Mus musculus (domesticus)*
c. *Sigmodon hispidus*
d. *Rattus exulans*
e. *Rattus rattus*
f. *Peromyscus maniculatus*
g. *Rattus norvegicus*
h. *Citellus tridecemlineatus*
i. *Microtus pennsylvanicus*
j. *Tamias*

42. Thirteen-lined ground
squirrel _____
43. Meadow vole _____
44. Pine mouse _____

k. *Pitymys pinetorum*
l. *Thomomys, Geomys* and *Cratogeomys*
m. *Sciurus*

Answers for Column A: e, d, b, j, m, g, a, c, f, l, h, i, k.

True or False

45. In some species of rodents, the sweat glands are confined to the soles of the feet or the skin between the toes.

    True

46. Rodents have guard hairs that are large and coarse.

    True

47. Rodents have under-hair or under-fur to keep them warm.

    True

48. Rodent incisors are chisel-shaped.

    True—This shape is created by the presence of enamel on the front surface of the teeth only. The backs of their teeth wear down more rapidly.

49. Some rodents are capable of hibernating.

    True

50. The hamster is the only rodent that increases its concentration of blood sugar in dormancy.

    True

51. Commensal rodent young are altricial. (They are born blind and helpless, and they stay in the nest for some time after birth.)

    True

4   VERTEBRATE PEST HANDBOOK

52. Rodents are mammals.

    True

53. Moles are rodents.

    False—They belong to the order Insectivora. They feed on insects, earthworms, other vertebrates and invertebrates.

54. Shrews are rodents.

    False—They belong to the order Insectivora. They feed on insects, mice and invertebrates.

55. When the means of transportation in urban areas changed from horses to automobiles, the mouse and rat populations in many of those areas declined.

    True—Before automobiles, we used horses, which required stables and provided more rodent food and harborage. People living in slum areas, however, would probably argue that there are more rodent harborages today.

Answer the Following

56. Indicate with a check which of the following arthropods are commonly known as rodent parasites.

    a. Thrips _____   e. Mites ✓
    b. Fleas ✓   f. Mosquitoes _____
    c. Lice ✓   g. Bed bugs _____
    d. Ticks ✓

Answers to above: You should have checked b, c, d and e.

57. Rank the three major domestic rodents in relation to the size of their droppings from largest to smallest.

    Norway rat 3/4 inch   1
    Roof rat 1/2 inch   2
    House mouse 1/8 inch to 1/4 inch   3

Rodents in General

58. How does the shape of mouse droppings differ from that of rats?

    Norway rat droppings are blunt compared to mouse droppings, which are distinctly pointed on both ends. Roof rat droppings are pointed on at least one end.

59. How can you tell if the rodent feces are fresh?

    They glisten and are soft. Old droppings are hard and dusty.

60. How can a rat gnaw on wires, wood and other hard objects without swallowing them?

    There is a space behind the incisors called a diastema. The rodent seals the area behind it and prevents any of the gnawed material from entering the gut.

61. Why do rodents seem to favor chewing on wire?

    It is twig-sized. In the wild, rodents feed on weeds and use them for nesting material.

62. What information can excreted urine transmit to other rodents?

    (a) whether species is foreign to it; (b) family or community member; (c) male or female; (d) whether in heat; and (e) whether dominant male.

63. What is a Florida water rat?

    This rodent, *Neofiber alleni*, is confined to Florida and the Okefenokee swamp of southeast Georgia. It is sometimes called the "round-tailed muskrat" because it looks like a young muskrat with a round tail.

64. What are rice rats?

    These are rodents belonging to the genus *Oryzomys*. They are found in the eastern United States, from eastern Kansas and Oklahoma across to New Jersey and south to Florida. They are also found in Central America.

65. What is a wood rat?

    Rodents belonging to the genus *Neotoma* are wood rats. They are also called pack rats, trade rats, mountain rats, brush rats and cave rats. These rodents are found in many areas of the southern and southwestern United States.

True or False

66. There are some rodents that weigh more than 60 pounds.

    True—A South American rodent called a capybara can weigh that much and more.

67. Rats originally came from Asia and spread west.

    True

68. The word *rodent* means "to gnaw."

    True

69. Rats can apply 8,000 pounds of pressure per square inch with their teeth.

    True

70. Rodents do not have to gnaw on objects to keep their incisors from growing too long, provided their upper and lower teeth can hit each other.

    True—At night, you can sometimes hear rats chipping their teeth by grinding them together, instead of having to gnaw on things.

71. The space between the front incisors and the rest of the teeth is called a diastema.

    True

72. Because of this space, rodents can draw in their mouths to plug their throats and can chew on wood, wire and other objects without swallowing them.

    True

73. Rodents spend up to 20 percent of their waking time grooming.

    True

74. Rats exhibit social grooming. They groom each other.

    True

75. Male mice can make a nest and often take over a nest made by a female mouse. **True**

76. Urine smells may influence whether or not a rodent will enter a bait station or be caught in a trap. **True**

77. Food items marked with a rodent's feces can be favored over food with no rodent droppings. **True**

78. Rats and mice will eat their own feces. **True**—They do this on occasion, probably to recover needed nutrients and beneficial bacteria.

79. The house mouse and Norway rat originally came from Asia. **True**

CHAPTER 2
Norway Rats
(Rattus norvegicus)

Answer the Following

80. Does the Norway rat come from Norway?

    No. The animal probably originated in Asia. It was given the name "Norway rat" by the Swedes in the 18th century, at a time when there was much rivalry between the two countries.

81. What are some other names for the Norway rat?

    It may be known as the brown, wharf, sewer, water, common or barn rat.

True or False

82. Norway rats living in cold buildings may have a thicker coat than rats living in warm buildings.   **True**

83. Newborn rats are blind and hairless.   **True**

84. Rats care for their young until the offspring are capable of being on their own.   **True**—If there is no undue stress on them, such as a food shortage.

85. Norway rats can interbreed with roof rats.   **False**

86. Norway rats can interbreed with mice.   **False**

Norway Rats (Rattus norvegicus)

87. Some Norway rats live in forests. True

88. Norway rats are good climbers. True

89. Norway rats are good swimmers. True

90. Norway rats can go three to four days without eating before weakness sets in. True

91. Norway rats rapidly decline due to weakness if they go without water for between one to two days. True

92. Rats can walk across telephone wires. **True**—Their tails can help them balance.

93. Rats are capable of hearing sounds that humans cannot hear. **True**—They can hear high-frequency sounds beyond the range of human hearing. These ultrasounds are important in rat communication. Newborn rat pups make ultrasounds that tell the mother rat where they are without alerting other animals.

94. The fecal droppings of a Norway rat are pointed at both ends. **False**—Pointed ends are characteristic of roof rat droppings. Norway rat droppings are blunt at the ends.

95. Rats travel along regular runways and are creatures of habit. True

96. Rats have highly developed senses and can learn from their experiences.   **True**

97. Rats sometimes leave "grease" marks on walls from body secretions.   **True**

98. A one-pound adult Norway rat is considered a large rat.   **True**

99. Rats eat about one-half pound of food every day.   **False**—They eat about one to three ounces of food per day.

100. Rats contaminate much more than they eat.   **True**—From urine, hair, feces, parasites and disease organisms.

101. Rodents lick their bodies with their tongues and swallow their own hair.   **True**

102. Rodent fecal pellets contain rodent hair.   **True**

103. There are albino Norway rats.   **True**—They were developed for medical research. Sometimes albino rats occur naturally in the wild, or develop from pet rats that escape.

104. Rats are more active at night.   **True**

105. Rats rarely live more than one year in the wild.   **True**

Norway Rats (Rattus norvegicus) 11

106. If a rat loses its tail, it can grow a new one.
False

107. If a rat gets its foot caught in a trap, it can chew its own leg off.
True

108. Rats are colorblind.
True

109. Rats cannot vomit.
True—This helps account for the use of zinc phosphide, an emetic that causes vomiting except in rodents and birds.

110. With a running start, Norway rats can jump three feet high.
True

111. When rats fight each other, they sometimes stand on their hind legs and box with their front legs.
True

112. Norway rats use their feet for digging.
True—They also use their noses and teeth.

113. Norway rats usually nest and live beneath the ground.
True

114. A young rat can squeeze through a hole the size of a quarter.
True

115. A young rat's head and feet are large in comparison to its body.
True—This is one way to distinguish young rats from mice.

116. Rats can lick the dew off grass to get enough liquid to survive.
True

117. Rats sometimes leave marks on dusty floors from dragging their tails.
True

118. The dominant male rats are responsible for successful breeding of 70 to 80 percent of the female population.
True

119. Rats have been known to swim through pipes into toilet bowls as a means of entering a house.
True

120. Some rats will only eat one type of food, and shun the most appetizing rodent baits.
True—So if one bait is not working, another type should be tried.

121. Liquid bait stations work best when the rodents' water supply is scarce or lacking.
True

122. Rats will be seen during the day if there is a shortage of food.
True—High populations or disturbance may also be the cause of this.

123. Norway rats vary in color from white to brown to black, and numerous shades in between.
True

124. Rats can cause fires.
True—Approximately five to 25 percent of fires of unknown origin are believed to have been started by pest rodents.

Norway Rats (Rattus norvegicus)

125. Dead rats should not be picked up with bare hands.

    True—Diseases and parasites pose a health hazard. Use gloves and put the carcasses in a plastic bag for burial, incineration or disposal.

126. Rats will fight each other and kill the weaker members of their group if there is a shortage of food or harborage.

    True

127. Rats can jump eight feet from one rooftop to another.

    True

128. Rats live in colonies.

    True

129. Rats can gnaw through galvanized metal.

    True—Their tooth enamel is harder than iron.

130. Rats can gnaw through cement.

    True—When too thin, "green" or improperly cured.

131. Rats have a sex drive.

    True—Females are in heat about every four days and can breed year-round.

132. One female rat will mate with several male rats.

    True

133. Rats are ready to breed when they are one month old.

    False—They breed when they are two to three months old.

134. Rats often stand in their food while feeding.

    True

135. Fifty-five gallon garbage cans should not be used for refuse containers.

True—They are too heavy when filled, encouraging less frequent attention and attracting rodents.

136. If need be, Norway rats have a home range radius of one-quarter mile or more.

True—When their habitat is disturbed or young rats are forced from the colony.

137. Norway rats prefer to remain within a 100- to 150-foot radius of their colonies.

True—When food, water and shelter are available.

138. Extremely hot and extremely cold weather decrease rodent reproduction.

True

139. Adult male rats fight to establish relative rank in a social order.

True

140. Rats exhibit cannibalism when available food is depleted.

True

141. Rats will abandon their newborn pups if there is a scarcity of food.

True—And may eat their offspring.

142. Rats build a definite nest, using bits of paper, rags, straw, etc.

True

143. Young rats build nests more readily at temperatures below 65 to 70 degrees Fahrenheit.

True

144. Adult rats build nests regardless of temperature.

True

145. Rats will move their young from one nest to another if disturbed.

True

Norway Rats (Rattus norvegicus)

146. Fresh rat droppings are soft, shiny and dark. — **True**

147. Rodent droppings harden within a few days. — **True**

148. Rats can scoop up water with their front paws. — **True**

149. Rats can readily detect movements. — **True**

150. Rats have poor distance eyesight. — **True**

151. Rats' tails look "naked." — **True**—But they have fine hairs.

152. Rats can swim for long periods without drowning. — **True**—They have been known to tread water in laboratory tests for up to three days.

153. Rats can push the lid off a can if it is not tight-fitting. — **True**

154. Brown rats are flat-footed. — **True**

155. Hairs shed by rodents can become airborne contaminants. — **True**

156. Rodent hairs often carry bacteria. — **True**

157. Rats can learn to avoid baits or traps if used the same way over a period of time. — **True**

158. Rats can develop food preferences during the first month of their lives and maintain those preferences. — **True**

159. Some species of tapeworms and nematodes have been associated with rats.
True

160. Every adult female rat is equally likely to become pregnant.
False—Only those females high enough in the social order of the colony are likely to have the opportunity to mate.

161. Female rats dominate the colony.
False—There is a definite social order, but it is headed by male rats.

162. Clear vision for rats is restricted to several feet, but movement is easy for them to detect.
True

163. Rats have better vision than mice.
True

164. Norway rats are native to the United States.
False—They came over on ships with the early settlers around 1775.

165. Rats originated in Asia.
True

166. In the laboratory, rats have lived for more than three years.
True—But this would be unusual in the wild.

167. In one year, under practical field conditions, one pair of rats can produce from 60 to 70 young.
True

168. Some laboratory rats can reach three pounds or more in weight.
False—A two-pound rat would be exceptional.

169. A rat can jump a horizontal distance of 18 feet.
True

170. Rats burrowing in embankments can cause floods.
True

Norway Rats (Rattus norvegicus)

171. There are an equal number of toes on a rat's front and back feet.

    False—There are four toes on the front feet and five toes on the back.

172. Female rats do most of the burrowing.

    True

173. The Norway rat is just one species of rodent. There are more than 1,000 different species of rodents in the world.

    True

174. Female rats can come into heat (be receptive to mating) every four or five days.

    True—If they are not already pregnant.

175. A rat can climb a rusty vertical pipe with a one-and-one-half to four-inch inside diameter, and can climb a rusty pipe with an outside diameter of one to three inches.

    True

176. Rodent control is a people problem, whereby the people have to be educated on principles of sanitation and rodent stoppage.

    True

177. Adult male rats are able to breed all year, but females have peak months for reproduction.

    True—Female rats can breed year-round in warmer climates, but have breeding peaks in the spring and fall in more temperate areas.

178. Some Norway rats develop specific food preferences.

    True—Some authorities believe it exists in at least some rat populations.

179. The initial reaction to new bait stations may be a decrease in eating the familiar food inside.

    True—Rats may avoid new objects for several days.

180. When baiting for rats, you should try to be quiet. The more noise you make, the longer it takes the rats to accept your bait.

True—Except in disturbed and noisy environments, where rats are more used to noise.

181. Rats can make nests in planters.

True—The soil is the rat's natural habitat. Remaining hidden in planters during the day, they come out at night to feed.

Fill in the Blank

182. Latin name for the Norway rat _____.

(*Rattus norvegicus*)

183. Another name for the rat's highly sensitive whiskers _____.

(vibrissae)

184. Length of time rat young stay in the nest _____.

(two to three weeks)

185. Area where rats originally evolved from _____.

(Asia)

186. Age at which rats start to breed _____.

(two to three months)

187. Number of rat litters per year _____.

(five to seven)

188. Gestation period of the rat (length of time pregnant) _____.

(about 28 days)

189. Average life span of a rat in the wild _____.

(six to eight months)

190. Percentage of adult female rats in the colony usually pregnant at any one time _____.

(20 to 30 percent)

Norway Rats (Rattus norvegicus)

Answer the Following

191. How deep can rats burrow?

    Up to six feet in some soils. But normally, their burrows are only eight to 18 inches below the surface of the ground.

192. Can rats gnaw through lead pipes?

    Yes, their teeth are harder than iron.

193. When disturbed, how far can rats travel on their own feet within a week?

    About four miles or more. If they hop on a railroad car or in a semitrailer truck, they can go across the entire country in a week!

194. How strong are a rat's teeth?

    By giving a few familiar items a hardness index and comparing them to a rat's teeth, you can get some idea why rats can chew through so many things.

Material	Hardness index
Rat tooth enamel	5.5
Iron	4.0
Copper	2.5–3.0
Aluminum	2.0

195. Must rats continually gnaw to keep their teeth from growing too long?

    No, they can sharpen their own teeth by the way they fit together if their teeth and jaws are in normal condition. If a tooth starts to grow crooked, it can enter the roof of the mouth and penetrate the brain, killing the rat.

196. By looking at a rat, can I tell if it is a subordinate ("omega") rat?

    Generally, these rodents have been beaten by the stronger alpha and beta rats. The omega rat may have more scars on its body, and its tail and ears may have been chewed.

197. How often do rats eat?

    Rats prefer to feed at night. Some experts believe that they feed twice during the night; once shortly after dark and again in the early morning. From experience, let us just say that rats eat when they are hungry. They prefer to eat in one place and differ from mice, which will eat many times in one evening.

198. How can you determine how many rats are present in an area?

This is very difficult. You begin by looking for rodents and rodent signs, including burrows, trails, gnaw marks, hairs, urine, tracks and odor. The use of a black light for urine and hair, and chalk dust for tracks, will help in determining the relative number of rats present. You can also use a monitoring bait, such as Census® bait. Rats will eat one to three ounces of food each, per day. They will stay hidden and show few signs where people are most active, which can mislead estimations.

A rough guide to determining the number of rats present has been developed by the U.S. Army:

DEGREE OF INFESTATION (NUMBER OF RATS PRESENT)

Rat Signs Observed	Light (1–20)	Medium (21–50)	Heavy (Over 50)
Tracks	Few to moderate; usually all of one size	Moderate to many; usually of two or more distinct sizes	Many; two or more sizes, slick runways in dusty areas
Fresh droppings	No or few groups observed, generally all of same size	Some always observable in two to six or eight areas; usually two distinct sizes	Many; usually of several sizes, small to large and in at least six locations
Active runs	None or few and relatively indistinct	Several distinct; one or more indicating heavy travel	Many; more than one heavily traveled run
Fresh gnawing	None to few instances nightly	Usually several instances nightly	Many instances nightly
Live rats seen	None by day, except on cleanup and harborage removal	None to two by day unless harborage is opened up	Often one or more seen by quite close observation, even in daytime

Taken from Army Medical Service School
Department of Professional Sciences
Preventive Medicine Branch
M-708-5 Code 046

Norway Rats (Rattus norvegicus)   21

199. Where should one look for rats in the back yard of a private home or apartment house?

    Rats harbor in a number of locations. Check the most common locations:

    1. Under doghouses
    2. Near bird feeders
    3. Garbage cans, sheds and recycling material
    4. Compost piles
    5. Vegetable gardens
    6. Debris buildup (lumber, tires, cut brush, etc.)
    7. Wood piles stored directly on the ground
    8. High grass and weeds
    9. Near or under landscaping timbers and railroad ties
    10. On a hill, check for burrows, especially in ivy, juniper or other ground cover
    11. Along sidewalks and walkways
    12. At the base of shrubs
    13. Properties next door and across the street, especially if they show heavy vegetation or potential harborage

CHAPTER 3
Roof Rats
(Rattus rattus)

True or False

200. Roof rats are smaller than mice.
False—Roof rats are considerably larger than house mice.

201. Norway rats are larger than roof rats.
True—But roof rats have longer tails.

202. In the United States, roof rats are more prevalent in the South, Gulf Coast and West Coast than in the North.
True

203. Roof rats are excellent climbers.
True

204. Roof rats are also called black rats.
True—Although they may be black, gray or brown.

205. "Ship rat" is another name for a roof rat.
True

206. Roof rats have long, prominent guard hairs.
True

207. The ears of a roof rat are smaller in proportion to its body than are the ears of a Norway rat.
False—The ears on a roof rat are more prominent than the ears of a Norway rat.

Roof Rats (Rattus rattus)

208. Roof rat droppings are spindle-shaped and more pointed at the ends. — **True**

209. These animals have a wider home range than house mice. — **True**—Roof rats have a home range of 100 to 150 feet, versus 10 to 30 feet for mice.

210. Roof rats can swim. — **True**

211. Roof rats have fewer young per litter than Norway rats. — **True**—Roof rats average six to seven per litter, versus eight to 10 per litter for Norway rats.

212. Roof rats have longer tails than Norway rats. — **True**

213. Roof rats build nests in attics, walls and trees. — **True**

214. Black rats walk on tiptoe. — **True**

215. Food preferences can be different for Norway rats and roof rats in the same building. — **True**

216. Roof rats prefer to feed on vegetables, fruit and cereal grains. — **True**

217. Roof rats are sometimes found in the Northeast and Midwest. — **True**—A few are sometimes brought into these areas on ships, trucks or trains. Usually, however, these individuals cannot sustain themselves to build a population in a colder climate.

218. Water is an essential daily requirement of roof rats.

True

219. It was the roof rat that was most associated with the plague (Black Death) in Europe in medieval times.

True

220. Roof rats live only in overhead areas.

False—Roof rats are capable of burrowing and are frequently found living underground, especially in tall grass or sugar cane fields.

221. The rodent most associated with ships is the roof rat.

True

222. The snout of a roof rat is much more pointed than that of a Norway rat.

True

223. The roof rat can live in sewers.

True

224. In California, the roof rat obtains both food and harborage by living inside Himalayan blackberry thickets and other vegetation.

True

225. Roof rats came to the United States before Norway rats.

True—Roof rats came on ships with the first settlers in 1620. Norway rats first arrived around 1775.

226. Roof rats will push Norway rats out.

False—Norway rats are larger and will drive roof rats out if they are competing for the same habitat.

Roof Rats (Rattus rattus) 25

227. Roof rats are colorblind. **True**—As are other rodents.

228. Roof rats like to feed on seeds. **True**

229. Roof rats can move up and down in pipes up to three inches in diameter. **True**

230. The Alexandrine rat is a roof rat. **True**

Match Them Up!

231. There are three color phases of the roof rat in the United States, although they interbreed. Match each color phase to the common name.

    Common Name
    a. Black rat _____
    b. Alexandrine rat _____
    c. Fruit rat _____

    Color Phase
    A. Black to slate gray
    B. Tawny above, grayish white below
    C. Tawny above, white to lemon-colored belly

    Answers: a = A, b = B, c = C.

Fill in the Blank

232. The Latin name for the roof rat _____. (*Rattus rattus*)

233. Average weight of adult roof rats in ounces _____. (eight)

234. Length of time required for a roof rat to reach sexual maturity in months _____. (two to three)

235. Approximate number of young roof rats weaned per year from a single female _____. (20)

236. Average litter size _____. (about six pups)

237. Average movement (home range) in feet _____. (100 to 150)

238. Average consumption of water per rat per day _____. (up to one ounce)

239. Type of transportation that roof rats are most commonly associated with _____. (ships)

Resources

240. **Can you direct me to specific references on roof rats?**

> Although there is much more literature on Norway rats, here is some material that stresses roof rats. Chapters on commensal rodents by Corrigan and Lund are readily available. Some of the other references may be located only through larger university libraries, the reprint collections of rodent experts, or from the authors themselves:
>
>> Brooks, J. E. "Roof Rats in Residential Areas—The Ecology of Invasion," *California Vector Views* 13 (1966): 69–74.
>>
>> Corrigan, R. M. Chapter 1 in *Handbook of Pest Control*, 8th ed. by A. Mallis. GIE Publishers, 1997.
>>
>> Dutson, Val J. "Use of the Himalayan Blackberry, *Rubus discolor*, by the Roof rat, *Rattus rattus* in California." *California Vector Views* 20 (1973): 59–68.
>>
>> Lund, M. Chapter 2 in *Rodent Pests and Their Control*, eds. A. P. Buckle and R. H. Smith. Cambridge: University Press, 1994.
>>
>> Marsh, R. E. and R. O. Baker. "Roof Rat Control—a Real Challenge." *Pest Management* 6, no. 8 (1987): 16–18, 20, 29.

Schwarz, E. and H. K. Schwarz. "A Monograph of the *Rattus rattus* group." *An. Esc. Nac. Ciene. Biol. Mexico* 14 (1967): 79–178.

Watson, J. S. "Some Observations on the Reproduction of *Rattus rattus.*" *Proc. Zool. Soc. London* 120 (1950): 1–12.

Yabe, T. "The Relation of Food Habits to the Ecological Distributions of the Norway Rat (*Rattus norvegicus*) and the Roof Rat (*Rattus rattus*)." *Japanese Journal of Ecology* 29 (1979): 235–244.

CHAPTER 4
House Mouse
(Mus musculus or Mus domesticus)

True or False

241. A mouse's fecal droppings can be colored to reflect what they have been eating.

    True—They may be brightly colored if, for example, they have been feeding on crayons or dyed rodenticide bait.

242. Mice can survive for long periods without having drinking water available.

    True—Mice can generally get what moisture they need from their food.

243. Mice are nibblers and they feed many times each day.

    True

244. A mouse's tail is "semi-naked."

    True

245. Mice are cannibalistic.

    True

246. Mice will not bite people.

    False—They will bite if handled.

247. Stress in their environment can affect mouse reproduction.

    True—Embryos may not form or mice may kill their young if stressed.

248. Mice have poor eyesight for seeing distances.

    True

House Mouse (Mus musculus or Mus domesticus)

249. Mice are colorblind. — **True**—They see shades of gray.

250. Male mice are generally found in the nest. — **False**—Females and young will commonly be found in the nest.

251. Male mice stake out territories with one male mouse per area. — **True**

252. A single mouse produces 50 or more droppings each day. — **True**

253. There are five to eight mice per litter. — **True**

254. Mice can be omnivorous (eating both plant and animal materials). — **True**

255. Many populations of mice prefer cereal grains over meat. — **True**

256. Newborn mice weigh about three one-hundredths of an ounce. — **True**

257. Once mice are accustomed to eating one type of food, it often becomes difficult to get them to alter their feeding habits. — **True**

258. Mice are born blind, pink and hairless. — **True**

259. A year is a long life for a mouse, but they can live somewhat longer on occasion. — **True**

260. Hearing is better developed in mice than vision. — **True**

261. There are albino mice. — **True**—They are used in research and sometimes occur in the wild.

262. A young mouse could easily squeeze through a hole the diameter of a dime. — **True**

263. Mice are smarter than rats. — **False**

264. Mice and rats can live in the same structure. — **True**—But they usually establish different territories.

265. The odor of mouse urine is very pungent to humans. — **True**

266. Mice have poor vision and cannot see clearly beyond about six inches. — **True**

267. House mice eat their body weight every 24 hours. — **False**—About 10 to 15 percent of their weight, or one-half ounce, per night.

268. All mice are house mice. — **False**—There are many kinds of mice.

269. House mice are native to the United States. — **False**—They originated in central Asia.

270. Approximately 50 percent of all mouse droppings contain *Salmonella* bacteria. — **False**—Less than one percent contain the bacteria.

271. "Popcorning" is a term used by some to describe the way mice move when jumping over objects. — **True**

House Mouse (Mus musculus or Mus domesticus)

272. "Meta-populations" is a term that refers to mice that have spun off the main population.

    True

273. Mice can visit a food source 200 times or more in a single night, although they may eat only three to four grams during that time.

    True

274. House mice can travel up to 200 feet in their search for food.

    True—But ordinarily they choose to live in an area where the distances to food are short.

275. Not all mice are curious. Some are naturally shy of traps and glue boards when first installed.

    True

276. A house mouse home range can cover a quarter-acre radius when living outside.

    True—Although mice infesting buildings may not move from an area of 100 square feet or less if it contains abundant food and adequate harborage.

277. Two or three female house mice can generate several pounds of nesting material in a year's time.

    True

278. House mice can make their own metabolic water.

    True—But if mice are feeding on a high-protein diet such as pet food, they need additional water.

279. A house mouse can spend its entire life within a bag of dog food.

    True

280. Mice are most active at dusk and a few hours thereafter.

True

281. Mice grow up to become rats.

False—They are separate species.

282. There are some house mice in the United States that have been found to be resistant to older anticoagulant rodenticides like warfarin.

True—In San Francisco and elsewhere. This is not a problem with today's effective rodenticides, however.

283. House mice may not eat rodent bait the first night you put it out.

True—Even though mice are considered curious, unless food is a limiting factor, mice may gnaw on wax blocks but not eat them. This can be determined if many gnawing fines are found around the blocks.

284. Young house mice are easier to trap than adults.

True

285. When placing a Ketch-All® or other multiple catch trap near a wall, place it so the wind-up handle or rod projecting from the other end is facing the wall.

False—If pushed, it will hit the wall and cause the trap to unwind or jam.

286. On occasion, house mice will eat window sash putty and window glazing.

True—They may be attracted to the linseed oil present in some putty products.

287. Tests show that house mice are particularly fond of pineapple, prunes, gumdrops and peanut butter.

True—These make excellent baits.

House Mouse (Mus musculus or Mus domesticus)

288. House mice are generally not as suspicious as rats of new foods, but will easily switch from rodent baits back to other foods.

    True

289. The key to trapping house mice is using the best bait.

    False—The key to successful trapping of pest rodents is location, location, location. The trap should be placed in a location between where the mice currently nest and where they eat.

See sections on Rodent Diseases, Rodent Control, and Rodenticides for more details on mice.

Answer the Following

290. Are house mice native to the United States?

    No. In the 1520s, they arrived on ships with explorers who landed in Florida and Latin America. Additional explorers and settlers brought more house mice to other East Coast ports.

291. How many mice can live outdoors on an acre of land?

    Calculations indicate that 82,000 mice could survive on an acre of land with lush vegetation (containing abundant food and shelter).

292. What is an easy way to distinguish a young rat from a house mouse?

    Look at its feet. Mice have small, dainty feet, while rats have large feet.

293. How does a house mouse behave when nervous?

    It begins to wag its tail frantically. In such a heightened state, mice become even more alert to their surroundings. Stressed mice may only seek shelter and will not seek food until they feel more comfortable.

34 VERTEBRATE PEST HANDBOOK

294. How can mice totally avoid tracking powder, traps or glue boards when these are directly in the pathway where they run?

   The running mouse can jump over the control materials. Mice can jump more than two feet from a running start.

295. How can you prove that mice are active in specific areas?

   You can look for urine or droppings, or use a non-toxic monitoring tracking powder such as chalk, or a monitoring block such as Census. For highly sensitive accounts, a video camera may also be mounted in position and set to run continually from dusk to early morning to provide visual proof.

Fill in the Blank

296. Latin name for the house mouse _____.   (*Mus domesticus*; prior name was *Mus musculus*)

297. Name a human viral disease associated with house mice _____.   (lymphocytic choriomeningitis)

298. Normal home range distance covered by house mice _____.   (10 to 30 feet)

299. House mouse adult weight _____.   (approximately one-half to one ounce, or 14 to 28 grams)

300. Age at which house mice reach sexual maturity _____.   (one-and-one-half months)

301. Number of young that female house mice wean each year _____.   (30 to 35)

302. Species of mammal most closely associated with the house mouse _____.   (humans, *Homo sapiens*)

303. Size crack in wall or under door that an adult mouse can squeeze through _____. (three-eighths inch)

304. Size hole that house mouse can squeeze through (diameter) _____. (one-quarter inch, therefore close holes the size of a pencil)

305. Number of droppings (feces) one pair of mice can deposit in six months _____. (18,000 droppings)

306. Fungus associated with house mice that can be transmitted to people and cause bald patches on the head _____. (*Favus*)

Answer the Following

307. Are histoplasmosis, rat-bite fever and tularemia associated with house mice?

    Yes.

308. How closely related are house mice and Norway rats?

    Both animals belong to the order Rodentia and the family Muridae; however, mice are in the genus *Mus* and Norway rats are in the genus *Rattus*.

309. What does *"Mus musculus"* mean?

    It comes from the Latin and means "little thief." This was the scientific name for many years, but it was later changed to *Mus domesticus*.

310. Will mice eat cockroaches off a glue board?

    Yes, although the legs of the cockroaches may be left in the glue.

311. Do pregnant female house mice go into a "nesting frenzy" prior to giving birth?

Yes.

312. For mice in an apartment, break down where to inspect—and possibly to treat—in each area.

Kitchen
1. Void area below kitchen cabinets
2. Hollow doors
3. Under and within the stove, including under the top burners
4. Behind and in the refrigerator compressor area
5. Wall voids
6. Suspended ceilings
7. In potted plants
8. In the broom closet
9. In accumulated items such as paper bags that have not been moved in some time

Bedroom
1. In wall voids
2. Under the bed, in debris or stored items
3. In soil of potted plants
4. In boxes and toys not used on a daily basis
5. Inside dressers, often in soft clothes
6. Inside filing cabinets, behind the bottom drawer
7. In back corners of closets
8. In suspended ceilings
9. Below floors where wires or pipes penetrate

Bathroom
1. Wall voids
2. In a hollow door
3. Under the sink cabinet
4. Where pipes penetrate the wall

Note: The attic and basement are also common locations where mice dwell, and should not be overlooked.

313. When inspecting a supermarket, where are the most likely locations that house mice will be found?
    1. Dry dog or cat food
    2. Bird seed
    3. Rice
    4. Dry cereal
    5. Candy and cookie section
    6. Cigarettes
    7. Locker areas

CHAPTER 5
Rodent Diseases

Human Diseases and Parasites Associated with Rodents
General Material

314. Indicate with the appropriate letter which of the following diseases and disease organisms are:

    A. Associated with rodents but transmitted by arthropods
    B. Associated with rodents and transmitted by rodents
    C. Not normally associated with rodents

315. Rat-bite fever _____

316. Rickettsialpox _____

317. Murine typhus fever _____

318. Leptospirosis _____

319. Tapeworms _____

320. Hantavirus _____

321. Plague _____

322. Lyme disease _____

323. Trichinosis _____

324. Malaria _____

325. Lymphocytic choriomeningitis _____

326. Salmonellosis _____

Answers to the above: B, A, A, B, B, B, A, A, B, C, B, B

38

Rodent Diseases 39

True or False

327. Ectoparasites on rodents should be killed before eliminating the rodent population.

    True—Otherwise, they will leave their dead host and seek a new host, which may be a pet or human.

328. Rats carry rabies in the United States.

    False—To date, there are no authentic cases documenting this.

329. Rat-borne diseases have killed more human beings than all the wars in history.

    True

330. Rats can carry lice, fleas and mites.

    True

Diseases Transmitted by the Rodents Themselves
Hantavirus

Questions 331–340 are adapted from *Healthbeat*, a newsletter published by the Illinois Department of Public Health in September, 1994.

331. What are hantaviruses?

    Hantaviruses are a group of viruses found in wild rodents. While they do not produce disease in these rodent hosts, hantaviruses can cause illness in humans. The viruses are so named because they were first isolated in the laboratory from striped field mice captured near Korea's Hantaan River. For many years, hantaviruses have been known to cause illnesses in other areas of the world outside the United States.

    In 1993, a previously unknown species of the virus, which causes an illness different from other known hantavirus infections, was identified in the southwestern United States. The form of the virus that causes illness and death in the United States is called *Sin Nombré* Virus (or "no name") because authorities could not agree on a better designation.

332. How is a person infected with a hantavirus?

Humans contract a hantavirus infection by breathing dust contaminated by the urine, saliva or feces of an infected rodent. Infection also may occur if contaminated material or dust gets into broken skin or a mucous membrane, such as the eye. Ingesting food or water tainted by an infected rodent may also cause illness. Hantaviruses also can be transmitted by the bite of an infected rodent. Person-to-person transmission has not been demonstrated.

333. What are the symptoms?

The most recently identified hantavirus can affect the lungs, so the illness has been named Hantavirus Pulmonary Syndrome, or HPS. Some types of pneumonia and common respiratory viruses (like influenza virus) can mimic the early symptoms of this hantavirus but, fortunately, HPS is rare. HPS's symptoms, which may develop between five and 42 days after exposure to the virus, include fever, headache, stomach pain, muscle aches, cough, and nausea and/or vomiting. If a person experiences flu-like symptoms followed by shortness of breath, he or she should contact a physician.

334. Who is most likely to get HPS?

Cases are most likely to occur in rural areas where the deer mouse, which appears to be the main source of the virus in the United States, primarily lives. Buildings, barns, garages or areas where rubbish or wood piles exist can serve as potential settings of hantavirus infection if such sites are inhabited by infected rodents and if conditions that are favorable for transmission (dry, dusty areas contaminated with rodent excreta) exist.

335. Is there a treatment for HPS?

HPS can be a serious, life-threatening illness. Treatment with ribavirin, a drug used with other hantaviruses, is being studied but has not yet proven effective. Support for patients with HPS is given in an intensive care unit, where fluids and blood pressure are maintained and mechanical ventilation with oxygen may be necessary.

Rodent Diseases   41

336. **What can I do to prevent rodent infestation in a home or building?**

    To keep rodents out of a building, you must create an environment that does not attract them. Deny rodents food, water, nesting sites and entry to the building:
    1. **Reduce the availability of food and water.** Keep your kitchen clean. Store human and pet food in tightly closed containers. Keep food scraps and garbage in rodent-proof metal or thick plastic containers with tight-fitting lids. Store bulk animal food at least 100 feet from the home in containers with tight-fitting lids. Do not allow pet or animal food to sit out. Repair leaky faucets that may provide water to rodents.
    2. **Eliminate nesting sites near the building.** Keep your lawn mowed; tall grass and weeds make an excellent habitat for rodents. When possible, follow the "100-foot rule"; plant gardens and place wood piles, compost heaps, feed bins and trash cans at least 100 feet from the home. Wood piles should be at least a foot off the ground. Haul away trash, abandoned vehicles, discarded tires and other items that could serve as rodent nesting sites. Place three inches of gravel under the base of mobile homes to discourage burrowing by rodents.
    3. **Seal the building.** Identify all possible sites of rodent entry. A mouse can fit through a hole slightly larger than one-quarter inch. Use steel screen, sheet metal, galvanized hardware cloth, caulk or weather stripping to seal holes or gaps along the edges of windows, entry doors and garage doors. Check places where pipes and electrical wiring enter the house and seal openings with steel wool.

337. **If I have a rodent problem in my home, what can I do to eliminate the infestation? Should I set out traps?**

    First, remove the three things required for survival: food, water, and places to hide and nest. Second, if rodents are present, set out snap traps, not cage traps, and be sure to follow the manufacturer's recommendations. (Peanut butter is excellent bait.) Continue trapping for at least two days after the last rodent is trapped. Third, maintain a rodent-free building by correcting conditions that attract rodents. Trapping is useless if the procedures to prevent reinfestations are not followed.

"Building out" rodents and trapping are the most effective control methods. Rodenticides should be used only to supplement these methods. Read and follow rodenticide label instructions. All rodenticides carry warnings that they be placed in tamper-proof bait boxes or in locations not accessible to children, pets and other domestic animals and wildlife. Fleas or mites, which can be a problem if there is a large infestation of rodents, are best controlled with an insecticide labeled for flea or mite control.

338. **What should I do if I find a trapped, poisoned or dead rodent in a house or barn?**

Wear intact rubber or plastic gloves when removing dead rodents and when cleaning or disinfecting items or areas contaminated by rodents. Soak or spray dead rodents with a disinfecting solution (see information that follows) until thoroughly wet and place in a plastic bag. The bag should then be placed in a second bag and tightly sealed. Dispose of rodents in trash containers with tight-fitting lids or by incineration. After handling rodents, resetting traps and cleaning contaminated objects or areas, thoroughly wash gloved hands in a general household disinfectant or in soap and hot water. Then remove gloves and thoroughly wash your hands with soap and warm water.

339. **What type of disinfectant should I use?**

The hantavirus is destroyed by detergents and readily available disinfectants such as diluted household bleach or products containing phenol (e.g., Lysol®). Be sure that what you choose is compatible with the item, object or area to be cleaned and disinfected.

A bleach/water solution (at least 1-1/2 cups of household bleach per gallon of water) destroys the virus when the item, object or area is thoroughly wet or is saturated with the solution during cleaning and disinfecting.

Products containing phenol will also destroy the virus when the item, object or area is thoroughly wet or is saturated with the solution during cleaning and disinfecting. If using a product containing phenol, be sure to follow label directions for use and recommended amounts.

Rodent Diseases 43

Detergent/water solutions destroy the virus when the item, object or area is thoroughly wet or is saturated and allowed a minimum of five to 10 minutes contact time with the solution. Follow label directions for product use and recommended amounts of laundry or dishwashing detergents. Detergent/water solutions may be helpful when the item, object or area requires removal of dirt.

Do not vacuum or sweep rodent-contaminated areas before cleaning, mopping or spraying with a disinfectant. This could cause virus particles in the dust to be spread into the air.

340. **Is hantavirus restricted to unpopulated areas in the western United States?**

    No. Most cases have occurred in the Four Corners area of New Mexico, Arizona, Colorado and Utah. However, cases have also been reported in the Midwest and in the eastern states of New York, Pennsylvania, Rhode Island, West Virginia, Virginia, North Carolina and Florida.

341. **What rodents are associated with the disease?**

    Deer mice, cotton rats and wood rats are known to spread the disease.

342. **Are house mice and Norway rats a threat?**

    Not yet, but be cautious when near their droppings, nesting material or the actual rodent. Be sure to wear gloves when removing rodents from traps or picking up carcasses from poisoning.

343. **How many cases have there been in the United States?**

    Southwestern states saw an outbreak in 1993 that killed 15 people. By 1994, there were about 50 cases, and by 1997 the number of cases had grown to 175, with about 73 fatalities (about a 40 percent fatality rate).

344. **How long does the virus remain viable in rodent urine, feces and saliva?**

    From two to 16 days. Thereafter, the virus is no longer contagious.

345. Who has a high risk of becoming infected with the hantavirus?

Trappers, hikers, pest control technicians and other people working around the rodents are more at risk. However, people sweeping up their cabins or sheds, checking in old trunks in the attic, and sleeping in rooms above where mice were present have contracted the disease.

346. What are the most important things to remember about preventing hantavirus infections?

1. Avoid coming into contact with rodents and rodent burrows (be cautious around woodpiles, garbage dumps or other areas rodents may frequent; never pick up a rodent carcass with your bare hands).
2. Don't sweep out, vacuum or use cabins, sheds or other enclosed areas that may have been infested by rodents until they have been disinfected.
3. Avoid potentially contaminated surfaces or materials before disinfecting, including water supplies (or use bottled water).
4. Always keep an eye out for rodents—while at work, home or during recreational activities.

347. What does the Centers for Disease Control and Prevention (CDC) recommend for safe disposal of dead rodents and/or rodent droppings and decontamination of affected areas?

1. Wear latex or rubber gloves.
2. Thoroughly wet down (spray) dead rodents, traps, droppings and contaminated areas with a general household disinfectant (Lysol, bleach, ammonia, etc.).
3. Discard disinfectant-soaked rodents and rodent waste with cleaning materials into a plastic bag and seal it. Then place into a second plastic bag and seal. If possible, burn or bury the bag (or check with your local or state health department about other appropriate disposal methods).
4. Disinfect floors, countertops and other surfaces with a general household disinfectant.
5. Before removing the gloves, wash gloved hands in disinfectant and then in soap and water. Thoroughly wash hands with soap and water after removing the gloves.

Rodent Diseases 45

6. Disinfect all used traps or bait stations, and then place in position for continued rodent control efforts.
7. Professionals working in potentially infested areas should use high-energy particulate air (HEPA) filters. HEPA filters can filter out particles as small as 0.3 microns, effectively removing the disease virus from the air. Conventional respirator filters cannot remove these small viral particles.

348. For areas in or near structures, what does the CDC recommend for prevention of rodents that may carry hantaviruses?

    1. Eliminate possible rodent nesting sites such as junked cars, old tires and trash piles.
    2. Dispose of trash and clutter promptly.
    3. Wash dishes and cooking utensils immediately after use and clean up spilled food.
    4. Do not leave animal food and water in feeding dishes overnight.
    5. Cut grass, brush and dense shrubbery within 100 feet of the home.
    6. Rodent-proof structures by sealing small openings.
    7. Continually maintain traps indoors; use rodenticide in bait stations outside.

349. Does hantavirus kill its rodent vector, such as the deer mouse?

    No.

350. Can hantavirus be spread from person to person?

    No.

351. Can hantavirus be spread to people by insects?

    No.

352. Are deer mice infected equally with hantavirus in different areas?

    No. In a study in New York state, mice on Shelter Island showed a 48.5 percent positive rate, while Long Island had 22.4 percent positive. The lower Hudson region had a 17.4 percent positive rate, but the other areas in New York that were sampled were less than one percent.

353. If hantavirus is suspected, what is the best rodent control approach to take?

    Use glue boards and snap traps for indoor rodent control. Live traps and rodenticides are not recommended, because the rodent may die in an inaccessible area and could maintain the virus for some time. Rodenticides may be used outdoors in tamper-resistant bait stations.

354. Is there a CDC hotline to get more information about hantavirus and other disease questions?

    Yes, call 888/232-3228.

355. Is hantavirus an ribonucleic acid (RNA) type virus that can mutate more easily than a deoxyribonucleic acid (DNA) type virus?

    Yes, and this is a concern; the virus may mutate and allow more rodent species to act as vectors of the disease to people.

356. Are voles (*Microtus* species) known to carry a non-pathogenic (non-disease causing) form of the hantavirus?

    Yes, as with rodents overseas, some forms of the virus are not as dangerous to humans as the Sin Nombré virus.

357. What is the most important step to tell a homeowner to take to help prevent hantavirus infection?

    Keep wild rodents out of your home and structures through rodent-proofing.

358. With the least risk, what is the best way to capture deer mice that may be infected with the disease?

    Use a flat multiple-catch trap such as a Victor Tin Cat® or Stick-All® Mouse and Insect trap. Use a glue board inside to help hold the carcass and feces, and therefore reduce the chance of infection from airborne disease particles.

Leptospirosis

359. What is leptospirosis?

    Leptospirosis is a disease of domestic rodents. The actual disease organism is a spirochete. There are several species, including *Leptospira icterohaemorrhagiae* and *L. ballum*.

360. How does the spirochete pass from the rodent?

    Rodents transmit it through their urine. The spirochete harbors in the rodent's kidneys.

361. Is this disease called by another name?

    Yes, some also call it Weil's Disease and infectious jaundice. CDC literature indicates, however, that Weil's disease is a rare and severe form of leptospirosis, and not just another name for it.

362. What kinds of domestic rodents transmit leptospirosis?

    House mice, roof rats and Norway rats are all capable of transmitting the disease.

363. How do humans contract leptospirosis from a rat?

    Primarily by handling infected rodents, but also by drinking or swimming in contaminated water or coming in contact with moist soil on which an infected rodent has urinated.

364. Can humans become infected other than by rodents?

    Yes. Cattle, swine and dogs also can transmit the spirochete. Infected rats and mice can keep the disease prevalent in pets and livestock.

365. What symptoms are prevalent in humans if they contract this disease?

    Eye inflammation, fever, small hemorrhages, muscle pains and severe headaches are characteristic symptoms. Its effects can range from mild to severe liver damage to jaundice.

366. Does refrigerating contaminated food destroy the disease organism?

    No. Foods stored at 39 to 41 degrees Fahrenheit still maintain a virulent spirochete population. In fact, cold temperatures seem to prolong the life of the spirochetes in food.

Lymphocytic Choriomeningitis

367. What type of disease is lymphocytic choriomeningitis (LCM)?

    It is a viral disease associated with house mice and some other mammals.

368. How does the virus pass from the mouse?

    Nasal secretions, urine, feces and semen can contain the virus.

369. What symptoms are prevalent in humans if they contract this disease?

    Sometimes it behaves like a case of the flu: fever, muscle pain and headaches may occur. The disease may begin with an influenza-like attack and terminate with recovery, or, after a few days of remission, meningeal symptoms may suddenly appear. Patients with meningoencephalitis show drowsiness, disturbed deep reflexes, paralysis and anesthesia of the skin.

370. What type of domestic rodents transmit this disease?

    Primarily, house mice. However, hamsters have also been infected.

371. Does it affect house mice?

    Sometimes the virus kills the mouse. Some female mice can harbor the virus and transmit it to their offspring without any ill effects. This means that the mice serve as a "storage area" for the virus.

372. Can humans die from this disease?

    Most people recover within a few weeks; occasionally it can be fatal.

Rat-bite Fever

373. What is rat-bite fever?

    A disease caused by bacteria. The specific bacterium is called *Streptobacillus moniliformis*.

374. Are there other names for this disease?

    Yes, it is also called Haverhill fever.

375. Why is it called rat-bite fever?

    The bacteria harbor in the gums and teeth of the rats. A human receives the bacteria when bitten by the rat. The result is a high fever.

376. How many people get bitten by rats each year?

    Conservative estimates place rat-bite incidence in the United States at about 14,000 people per year.

377. What type of people get bitten most often?

    Children under 12 are bitten more often than adults. There are probably several reasons for this. Some children go to sleep with food particles on their faces and hands. The rodents may investigate the food particles and bite when startled by the movements of the child. Other times, children reach for the rat to play with it and are bitten on the fingers.

378. What part of the human do rats bite most often?

    The fingers and hands. This occurs when one reaches for or goes to brush away the animal.

379. Does everybody who gets bitten contract rat-bite fever?

    No. The number of cases of people contracting these bacteria via rat bites is very small compared to the number of people who are bitten.

380. What symptoms are prevalent in humans if they contract the disease?

    Three to 10 days after the rat bite, a primary lesion appears. There is swelling of the regional lymph nodes, a rash, arthritic symptoms and sharp fever peaks. If untreated, the mortality rate is about 10 percent.

381. Are there any secondary complications that can occur as a result of rat bites?

    Yes. There is always a danger of contracting tetanus (continuous contraction of muscles caused by toxin produced by bacteria) through an open wound.

Salmonellosis

382. What is salmonellosis?

    Food poisoning caused by salmonella bacteria. More than 30 different species can survive in humans.

383. Do rodents carry all types of salmonella?

    No. *Salmonella typhimurium* and *S. enteritidis* are the two major salmonella bacteria associated with rodents.

384. Is this disease very prevalent in rodents?

    It is thought that few rodents are infected with the food poisoning organisms.

385. What kinds of domestic rodents are associated with salmonella?

    Roof rats, Norway rats and house mice are all capable of transmitting the bacteria.

386. How is the bacteria transmitted to humans?

    Although it is possible for humans to contract salmonella from rodent feces, infection through this route is rare. Eating food left unrefrigerated or in unsanitary conditions is a much more common route of infection.

387. What symptoms are prevalent in humans if they contract this disease?

    Mild to severe intestinal disruptions are common. Death can occur in severe cases.

388. Is the bacteria harmful to the rodents?

    Yes. In some cases, it can even cause the death of the rodent.

Tapeworms

389. What are tapeworms?

    Tapeworms are flatworms belonging to a large group of animals called cestodes. The dwarf tapeworm, *Hymenolepis nana,* is associated with rodents and can be transmitted by rodents.

390. How does the tapeworm get from the rodent to the human?

    Dwarf tapeworms can be passed from the rodent to the human through the rodent feces. The eggs of the tapeworm are inside the feces.

391. What kind of rodents transmit tapeworms to humans?

Although not a common occurrence, the commensal rat and mouse may transmit the tapeworm.

Trichinosis

392. What is trichinosis?

This disease is caused by a round worm called *Trichinella spiralis*, most commonly thought of in association with eating infected pork.

393. How are rodents implicated?

Rodents contaminate each other as well as swine by eating infected feces. The roundworm can create a cyst form that remains viable for up to 27 days. Hogs can become infected by eating infected rat feces.

394. What types of domestic rodents transmit this roundworm?

Norway rats, roof rats and house mice are all capable of transmitting *Trichinella spiralis*.

Diseases Transmitted by Arthropod Parasites of Rodents

Lyme Disease

395. What is Lyme disease?

It is a spirochete infection that often begins with a characteristic rash. Later, it can adversely affect the joints, nervous system and heart.

396. What is the name of the spirochete causing Lyme disease?

Borrelia burgdorferi.

397. How can one contract Lyme disease?

The most likely method of contracting the disease is to be bitten by an infected tick.

398. Which ticks carry the spirochete?

Originally, it was believed that three main species were involved. These were the deer tick, *Ixodes dammini*, in the East; the black-legged tick, *Ixodes scapularis*, in the South; and the Western black-legged tick, *Ixodes pacificus*, in the West. In 1993, taxonomists lumped the East and South species as one and dropped the name *I. dammini*.

399. Where does the lone star tick, *Amblyomma americanum*, fit into the picture?

In the South, it is a possible vector but it is not a very efficient transmitter of the spirochete.

400. Are all ticks infected?

No. The percentage of ticks infected varies greatly from area to area. In some areas, such as parts of New York and Connecticut, the infection rate can be 20 percent or higher. Some reports claim 100 percent of the ticks in a given area can be infected, but this is rare. Therefore, not every tick that bites a person is infected.

401. Are the ticks born with the spirochete?

No. They must pick it up from the white-footed mouse or other vector.

402. Are humans the only ones to be affected by the spirochete?

No. Dogs are very susceptible. Cats, horses and cows can also contract the disease.

403. Can you get the spirochete other than through the tick?

On rare occasions, it can be transmitted via a blood transfusion. Mice can transmit it from mouse to mouse by licking infected urine. Cows are also known to get it in this manner.

404. Why is the tick called the deer tick?

In the East, adult ticks overwinter on deer. However, they also overwinter on raccoons, dogs and other mammals. Therefore, the name is not entirely appropriate.

Rodent Diseases 53

405. **Why is it called Lyme disease?**

The first time it was recognized in the United States was in Lyme, Conn. A group of children came down with a rare type of arthritis. The mothers of those children helped prod the doctors to find out why.

406. **How widespread is the disease?**

It has been reported in all 50 states, but is most prevalent where migrating birds settle. It is believed the disease has been spread by birds.

407. **Do people realize it when they are bitten by a tick?**

Most people do not know they have been bitten by a tick. The tick is very small and the bite is not painful. Less than 20 percent of people who have Lyme disease remember getting bitten by a tick.

408. **What are the early symptoms of Lyme disease?**

The Massachusetts Department of Public Health summarizes it as follows:

The first symptom of Lyme disease is usually a skin rash called erythema chronicum migrans (ECM), which occurs at the site of the tick bite. The actual tick may go unrecognized. The rash, which develops three to 32 days after the tick bite, begins as a small red area that gradually enlarges, often with partial clearing in the center of the lesion so that it resembles a doughnut or bull's eye. There may be multiple secondary lesions. The skin lesion is occasionally described as burning or itching. A small number of people with Lyme disease may not have the early skin rash, and symptoms may appear only in the later stages of the disease. Other skin signs include hives, redness of the cheeks and under the eyes (malar rash), and swelling of the eyelids with reddening of whites of the eyes. These skin signs may be accompanied by flulike symptoms such as fever, headache, stiff neck, sore and aching muscles and joints, fatigue, sore throat and swollen glands. If not treated, these symptoms may disappear on their own over a period of weeks; however, the rash may reoccur in about 50 percent of untreated people.

More serious problems may follow. If treated with appropriate antibiotics, the skin rash goes away within days, and complications may be avoided.

409. **What are the late symptoms of Lyme disease?**

There are three major organ systems involved in later stages of the disease—the joints, the nervous system and the heart. These can become affected weeks to months after the initial symptoms, although symptoms typically appear four to six weeks after the initial tick bite.

Symptoms in the joints occur in up to 60 percent of untreated people. This is an arthritis affecting the large joints (primarily the knee, elbow and wrist), which can move from joint to joint and can become chronic.

Neurological complications occur in 10 to 20 percent of infected people. The most common symptoms include severe headache and stiff neck (aseptic meningitis), facial paralysis (or other cranial nerve palsies), and weakness or pain of the chest or extremities, or both.

These symptoms can persist for weeks, often fluctuate in severity and may respond to intravenous antibiotics.

410. **Is there a peak season when you are most likely to get the spirochete?**

Yes. May through September is the peak time. Fortunately, deer hunters are not normally out in the woods during these months. It is possible, however, to get the disease at any time during the year.

411. **What is the life cycle of the deer tick?**

In winter, the tick needs to feed (deer are preferred) to lay eggs in the spring. The adults die after laying eggs. The eggs, deposited outdoors in the spring, hatch several weeks later. The larval ticks need to feed once during the summer, usually on mice, and they will develop into nymphs the following summer. The nymph feeds once during the summer on mice, dogs, deer or humans before maturing into an adult in the fall. Then the adult ticks attach themselves to a host, usually a white-tailed deer, where they mate. The female lays its eggs and the two-year cycle is repeated. In some instances, the life cycle may require three years.

Rodent Diseases   55

412. Must deer be present for the tick to complete its life cycle?

    No. In areas where deer are not present or have not been found for 20 years or more, the disease is still present and the ticks survive on alternative hosts.

413. Will deer ticks come after you?

    Compared to other ticks, they are very slow-moving and you have to come to them. By brushing up against them, you can pick them up.

414. What can an individual do to minimize the chance of getting Lyme disease?
    1. After walking in the woods, check for ticks.
    2. Wear a hat when walking in the woods. Ticks can drop from overhead tree branches, although more often they are attached to the tops of tall grass stalks.
    3. Avoid walking in an area where tall and short grass, or trees and fields, come together. This ecotone is favored by ticks because it is a natural corridor or pathway for animal activity.
    4. Keep grass cut low along fence lines.
    5. Wear repellents on your clothes. Products containing DEET or permethrin are effective materials.
    6. Do not feed birds on your property. The seeds attract mice, which attract ticks.
    7. When outdoors, dogs should be kept on a cement pad rather than on the ground. The heat from the sun on concrete will dehydrate and kill ticks.
    8. Keep garbage can lids on tight to deter raccoons and dogs from frequenting the area. These hosts carry the ticks.
    9. Rake leaf litter to decrease harborage areas for the ticks.
    10. Wear white socks when in the woods to better detect ticks.
    11. When playing golf, stay out of the woods and rough areas. Drop another ball.
    12. Wear light-colored clothes so you can see the ticks more easily.
    13. If you plan to walk in areas that are vulnerable to tick infestation, tuck your pants inside your socks and tape the area where they meet.

14. Check children for ticks when they come into the house. It usually takes up to four hours for the tick to start feeding, so there is time to get the tick off.
15. Encourage the vet to check your pets' blood to determine if they are carrying the spirochete.
16. Make people aware that having a cat outdoors may increase the chance of getting the disease.
17. Install a chimney screen to keep raccoons and squirrels away.

415. How long will a tick feed on a mouse?

    Only for about four days.

416. How many ticks will attach themselves to a single mouse?

    Hundreds have been found on a single mouse at one time.

417. In which life stages will a tick attack a mouse?

    The larval and nymphal stages.

418. In which life stage is the tick most likely to transmit the spirochete to humans?

    The nymphal stage.

419. Does the nymphal stage feed on animals other than humans?

    Yes, mice and domestic animals.

420. How long does the tick have to feed on a human to transmit the spirochete?

    We do not know for humans, but believe that it is several hours. Studies with other animals indicate 36 to 48 hours of feeding is necessary.

421. Does the tick cause pain when it is feeding?

    No. There is no pain at all. You do not feel a thing while they are feeding.

422. Which species of mouse is a main reservoir host for this spirochete?

    The white-footed mouse, *Peromyscus leucopus.*

Rodent Diseases 57

423. **Is the rodent most associated with the tick found throughout the United States?**

   No. In the northern West Coast, kangaroo rats and wood rats (pack rats) are the main carriers. In California, none of the white-footed mice had the spirochete.

424. **Are ticks born with the disease?**

   No. The tick eggs hatch and the larval ticks must seek out an infected white-footed mouse in order to become infected. They get the spirochete when feeding on the blood of infected mice.

425. **Do ticks infect animals in the same numbers year-round?**

   No. During the early spring, wood ticks seem more common. By late June through September, the deer ticks take over.

426. **Where do white-footed mice nest?**

   These rodents nest at ground level, but when the ground is very wet, they are known to move their litter and nest in trees at the base of squirrel nests.

427. **Where can I obtain more information on Lyme disease?**

   There are numerous references and associations that can help you. For more information, contact:

   1. Your physician.
   2. Your local health department.
   3. Your state health department.
   4. The Federal Centers for Disease Control and Prevention (CDC) 888/232-3228 or 404/332-4555.
   5. The Arthritis Foundation, Connecticut Chapter 860/563-1177.
   6. The Lyme Disease Foundation 800/886-5963.
   7. The American Lyme Disease Foundation, Inc. 800/876-5963.

428. **How effective is the test on humans to verify if you have Lyme disease?**

   Beginning in 1993, a test using joint fluid was proven very effective. Lyme disease was positively identified in 70 out of 73 arthritis patients. This work was done at Tufts University and the Mayo Clinic. Researchers are continuing to improve detection methods for the disease.

429. **Is there any correlation between having a cat and getting Lyme disease?**

    The *New England Journal of Medicine* published a paper indicating that there is a connection. Perhaps the cat picks up the tick and carries it into the structure. Once inside, the tick prefers the human over the cat.

430. **What is the best way to remove a tick?**

    Use tweezers and pull steadily but gently. There are special tick removal tweezers. They can be obtained by writing to:

    Instruments of Sweden, Inc.
    P.O. Box 10810
    Waterbury, Conn. 06726

431. **How long has Lyme disease been in the United States?**

    Historical records seem to indicate soldiers in the American Revolution on Long Island may have had the disease. We do not know why it remained obscure until the late 1970s.

432. **If it is necessary to send a tick for testing to see if it is carrying the disease, how do I send it?**

    Place the tick in a non-crushable container with a slightly dampened piece of paper towel. This is needed to prevent the tick from dehydrating. Do not soak the paper towel.

433. **Is the tick the only arthropod shown to carry the Lyme disease spirochete?**

    By far, the tick is the main vector. In experiments, house flies, deer flies, mosquitoes and fleas have been known to harbor the spirochete. Researchers are still investigating how successful these insects are in transmitting the spirochete in the field.

434. **Has the house mouse ever been associated with this disease?**

    In March 1989, in Georgia, the Lyme disease spirochete was found in the house mouse. The extent to which house mice are linked with this disease has not been established.

435. **What symptoms does a dog infected with Lyme disease exhibit?**

    Besides lethargy, the dog may appear arthritic. Check with a veterinarian for more detailed information.

Rodent Diseases

436. Does a cold winter reduce tick problems?

   No.

437. What biological agents are used to control ticks?

   Some people have tried to keep guinea hens to control ticks around their homes. In fact, the famous entertainers Christie Brinkley and Billy Joel, when living on Fire Island, kept these birds. Unfortunately, they proved ineffective, as Brinkley came down with the disease.

438. Can people get the disease a second time?

   Yes.

439. Is there a blood test for Lyme disease?

   Yes, but timing is important. You must wait a few weeks after being bitten for the test to show positive.

440. Is there a vaccine to prevent a person from contracting the disease?

   For dogs, a vaccine has been developed. The vaccine is about 80 percent effective. A human vaccine has been developed, but not for use for children four and under. A trial is under way in the younger population, four to 14 years old. For more information about Lyme disease vaccines, contact Dr. Allen Stere, New England Medical Center, 617/636-5951, or write to Department of Medicine, 750 Washington St., NEMC #406, Boston, Mass. 02111.

441. Are dogs capable of transmitting the Lyme disease spirochete to ticks?

   In 1984, the New York Medical College, in conjunction with the University of Rhode Island Center of Vector-Borne Disease, showed that dogs were good carriers of the disease. Infected beagles were able to transmit the disease to uninfected, immature ticks feeding on them.

442. Once you have contracted the spirochete, will you show positive in tests thereafter?

   Unfortunately, yes. This makes it difficult to determine if you have contracted the disease again.

443. Do some people test positive on the Lyme disease blood test but, in fact, do not have the disease?

   Yes, the test is not completely accurate.

444. Do some people test negative for the spirochete and, in fact, have the disease?

   Yes. A spinal tap can give a more positive identification in such situations.

445. Can you speculate on the reason there are fewer cases of Lyme disease in humans in the South compared to the Northeast?

   The major *Ixodes* tick associated with this disease is *Ixodes scapularis*. This tick feeds primarily on reptiles in the South (such as lizards). Therefore, they have less chance of picking up the spirochete than the same *Ixodes* ticks in the Northeast that feed on white-footed mice, the rodent that acts as a reservoir host of the Lyme disease organism.

446. Are incidences of human Lyme disease on the rise?

   Yes. Since 1982, more than 50,000 Lyme disease cases have been reported in the United States. Westchester County is an affluent suburban neighborhood north of New York City and a hotbed for this disease. There were 7,943 cases in 1990 and 9,344 cases in 1991. This occurred in spite of public notices and a media blitz to make people aware of the precautionary measures they should take. The worst areas are from Virginia north to Maine, and the Midwestern states of Indiana, Illinois, Michigan, Wisconsin, Minnesota and Iowa. In the West, the disease is found from California north to Washington, and also Arizona, New Mexico and Nevada.

447. Is the Northeast the major area where Lyme disease occurs?

   Yes. In 1994, Lyme disease hit record levels in the eastern United States. Maryland, New York, New Jersey, Connecticut, Rhode Island and Pennsylvania are the primary areas. There is some concern on the reporting system, however. In Connecticut, a study misdiagnosed 146 children as having Lyme disease. In actuality, they had sinus infections, viral infections or rheumatoid arthritis. The story is further complicated by a new disease called human granulocytic ehrlichiosis (HGE). (See questions 473–480.)

448. **Can you list any medical locations that specialize in Lyme disease?**

New York Medical College Lyme Disease Center, Valhalia, N.Y.; 914/493-8515.

Lyme Disease Center, Long Island Jewish Medical Center, New Hyde Park, N.Y.; 718/470-3480 for children, 718/470-7290 for adults.

Lyme Disease Resource Center, Jersey Shore Medical Center, Neptune, N.J.; 732/775-5500.

Lyme Disease Institute, University Hospital, Stony Brook, N.Y.; 516/444-3808.

449. **Is there an effective preventative vaccine for humans?**

Yes, one was recently developed.

450. **Is there a reference that can teach children about Lyme disease?**

The Biomedical Research for Animals on People (BIORAP) publication is a science newsletter put out by Connecticut United for Research Excellence (CURE). They provide an approach on how to educate children about Lyme disease. CURE can be reached at P.O. Box 5048, Wallingford, Conn. 06492-7548; 203/294-3521.

451. **Do people who are bitten by deer ticks automatically need antibiotics to prevent Lyme disease?**

No. Even in areas where the incidence is known, only one to two percent of people bitten become infected. Most infections can be successfully treated with antibiotics if caught early.

452. **Do some people receive early medication because of the severity of Lyme disease?**

Yes. Pregnant women may because of the risk to their unborn children.

453. Can chipmunks and squirrels help spread the Lyme disease organism to ticks?

Yes, but again, these rodents are probably four times less likely than the white-footed mouse to spread the disease by being hosts for ticks.

454. Must business property owners take precautions to protect workers on their property from contracting Lyme disease?

Yes. Four track workers employed by the Long Island Railroad were awarded more than $560,000 to compensate for pain and suffering, after they contracted the disease while working on the railroad.

455. Does using fences to "build out" deer help reduce deer ticks?

Yes. In 1994, researchers at the Connecticut Agricultural Experimental Station reported that when fencing off an eight-acre and an 18-acre parcel, the number of ticks found within the enclosures were generally reduced, with larval ticks reduced by 97 percent.

456. Why are carbon dioxide traps so ineffective in trapping deer ticks compared to other tick species?

It appears that deer ticks are not very mobile. Rather than actively seeking a host, they wait for one to come by. Deer ticks get on you when you brush against them. Therefore, tick drags are more effective in finding these arthropods.

457. How does one conduct a tick drag?

A tick drag can be conducted by fastening one edge of an old blanket or other coarse-weave fabric to a pole and dragging the blanket through and over vegetation in infected areas. A convenient size that can allow dragging in more confined areas would be about two feet wide and seven feet long. Weights attached or sewn into the trailing corners of the fabric will help keep the drag straight and in contact with vegetation. Ticks that are "questing" (perched on vegetation, waiting for a host to pass) will readily attach to the blanket. Ticks can then be killed by placing the blanket in a plastic bag in the hot sun.

Rodent Diseases

458. Is Lyme disease considered an occupational disease?

    The official designation of this disease as an occupational disease varies from state to state. In New York state, people contracting Lyme disease on the job are eligible for Worker's Compensation.

459. Where can preserved specimens of the Lyme disease vector tick and the spirochete disease agent be obtained?

    Excellent teaching aids, including specimens of these organisms, can be obtained from several biological supply houses, including:

    Ward's Natural Science Establishment, Inc.
    P.O. Box 92912
    Rochester, N.Y. 14692-9978

460. Are the deer ticks that carry Lyme disease native to the United States?

    No. It is believed that these ticks were transported on seabirds from Europe.

461. When were the deer ticks first discovered in the United States and where?

    They were first reported in 1926 on deer on an island off Cape Cod, Mass. The spread of the tick was not well-documented, but by 1969, the species was recorded in Wisconsin and elsewhere.

462. What is the role of birds in spreading this disease today?

    House wrens, robins and the common grackle are the primary vectors of Lyme disease. In one study done in Westchester, N.Y., American robins accounted for more than 70 percent of all the larval ticks on birds. Both robins and house wrens are reservoir hosts for the tick harboring the Lyme disease spirochete. Robins are migrating birds and, as such, can transport the spirochete to new locations.

463. What is the best time to apply pesticide around a home to achieve control of the vector tick?

    Studies in the Northeast show that one treatment around a home with an effective acaracide in May will control the nymphs, and that another application is necessary in October to control the adults. Making repeated applications between the May and October treatments is not necessary.

464. **If only a single pesticide application is made, which one is more important, the one in May or the one in October?**

The early summer (May) treatment is more important. This can reduce the risk of Lyme disease in an area by 70 to 90 percent.

465. **What are the most effective pesticides to use to control ticks?**

Chlorpyrifos and carbaryl granules (Dursban® and Sevin®) have been used in the past for wet spring-summer applications. Granular materials are more effective after rainfall or irrigation, but their residual effects may be limited. Liquid pesticides are effective in the drier fall weather or when applied before several days of dry spring weather. New pyrethroids are highly effective, including cypermethrin (Demon®, Cynoff®), lambda-cyhalothrin (Demand®) and deltamethrin (Suspend®). Microencapsulated products such as Demand can provide for increased weatherability in sunlight and longer residual effectiveness than emulsifiable concentrate or wettable powder formulations.

466. **What types of tests have been done to evaluate the control effectiveness of various pesticides?**

Individual manufacturers may have specific test data concerning their products and tick reduction efficacy. In a New York study involving several products, there was a 94 percent reduction in ticks with a single pesticide application of either chlorpyrifos, carbaryl or cyfluthrin. The control was measured using tick drags before, during and after the tick season on 160 residential lawns.

467. **Where do most deer ticks harbor?**

In a study conducted by Rutgers Cooperative Extension Service, 34 homes on lots of about one-half to one-and-one-half acres were studied. Most of the vector ticks were found to reside in the woods around the lots. The second most common area was between the woods and grass fields (ecotone). The lawns were the least desirable areas for ticks. In some cases, ticks were found on ornamental plantings next to home foundations.

Rodent Diseases 65

468. **What is the product Damminix®?**

    This product comprises cotton balls dipped in permethrin (a pyrethroid insecticide/acaricide), placed in cardboard tubes. While less effective than a perimeter treatment around homes that can cover lawns and leaf litter, Damminix yields some tick reduction and is useful in wooded areas or other locations that cannot be sprayed. Success varies in different situations.

469. **How does Damminix work?**

    Damminix works on the principle that deer mice, as with most rodents, regularly collect materials to use for nesting. Deer mice and white-footed mice take the treated cotton balls back to their harborage for nesting material. Ticks already on the mice are exposed to the pesticide, and those attracted to infested areas and nests also come in contact with the pesticide and die.

470. **How is Damminix placed?**

    The objective in using Damminix is to "flood" the property with the Damminix-treated placements. Follow the label directions, placing under shrubs and along fence lines and other prime habitats for deer mice.

471. **How effective is Damminix?**

    Effectiveness of this product is limited by several factors. The smaller the area treated, the fewer ticks can be controlled. So, for example, doing a single residential lot of one-half acre in the midst of a larger infested area will have limited results, and will leave many infected ticks presenting a potential risk to people coming and going from the area. Also, in areas where other rodents such as chipmunks may carry the infected ticks, there may be less use of the cotton for nesting, making the Damminix less effective than it is with deer mice and white-footed mice.

472. What steps can a pest control operator (PCO) take to help customers decrease the chance of contracting Lyme disease?

Educating customers about what they can do (a) for personal protection, and (b) to reduce the amount of favorable habitat for ticks and mice on their property is very important. Use brochures available from your local, state or federal health departments or government health agencies, or create one for your own company. A personal protection brochure should outline the nature of the disease and the life cycle of the tick vector. It should include information on how to dress to reduce the risk of acquiring ticks and how to check for and remove ticks on the body. Tick repellent applications are useful. Personal habits and behaviors rank next in importance, stressing areas to avoid if possible, such as infested woods and ecotones (areas where grass or fields and trees meet).

Property protection tips should follow. These may include: keeping grass near homes and harborage areas (fences, woods) cut low; limiting the feeding of birds and wildlife; rodent-proofing; using vent screening and chimney caps to prevent home access by vermin such as mice, squirrels and raccoons; reducing mulch and litter near homes, and storing trash and garbage properly in vermin-proof containers.

Homeowners should be encouraged to provide tick control for their pets using effective products, and have a vet determine if their dog or cat is infected with the disease spirochete. In infested areas, limiting pet access to outdoor areas will help prevent carrying ticks back to residents.

PCOs can offer services that include rodent-proofing and harborage reduction (removal of firewood to at least 100 feet from the home) as well as pesticide applications. Barrier or perimeter applications of pesticides for tick control can be effective when they are made around homes, vegetation and mulch/litter areas, as well as in yards or other areas where ticks may be present. Spring and early summer pesticide applications will be more effective in breaking the tick life cycle and eliminating newly hatched larvae before they have a chance to feed on rodents or other hosts.

Tick drags can be conducted by PCOs on a routine basis in areas where ticks have been reported, or around and near homes adjacent to heavily wooded or uninhabited areas. When done on a routine basis, tick drags can indicate tick incidence, treatment effectiveness and the need for further control efforts.

Human Granulocytic Ehrlichiosis (HGE)

473. What is human granulocytic ehrlichiosis?

This is another tick-borne disease caused by *Ehrlichia chaffecusis*.

474. Which ticks carry the disease?

Both dog ticks and deer ticks are vectors.

475. What are the symptoms of the disease?

High fever, severe headache, muscle aches, loss of appetite, nausea and vomiting appearing 10 to 14 days after the tick bite.

476. Can the disease be fatal?

Yes. Four people died from it in mid-1994.

477. Where has the disease been reported?

It has appeared in about 24 states, but is most prevalent in Wisconsin and Minnesota. There was also a case reported in Connecticut.

478. Is this a new disease?

It was first recognized in 1986. So far, no more than about 150 cases have been documented.

479. What is the treatment for the disease?

The antibiotics tetracycline and doxycycline are known to be effective.

480. Can a person have both Lyme disease and HGE?

Yes. Some doctors believe that about 10 percent of people with Lyme disease may also have HGE.

Murine Typhus Fever

481. **What is murine typhus?**

    This disease is caused by a rickettsia, a small organism slightly larger than a virus, but still able to live within a cell. The specific organism is *Rickettsia prowazeki* (var. *typhi*).

482. **Is murine typhus called by other names?**

    Yes, it is also called endemic typhus.

483. **How does the rickettsia get from the rodent to humans?**

    Various flea species, especially the Oriental rat flea, bite the infected rodent and pick up the rickettsia. The rickettsia is then in the flea, and, when the flea bites a person, it also leaves contaminated flea feces with the rickettsia near the puncture site. The human then unknowingly scratches or rubs the flea bite area, and the rickettsia is rubbed into the wound.

484. **What type of domestic rodents are most associated with this disease?**

    Roof rats and Norway rats.

485. **How prevalent is murine typhus in the United States?**

    This disease was more common before the advent of DDT, when about 5,000 people were stricken annually. Today, less than 100 cases occur each year.

486. **Where is the disease most prevalent in the United States?**

    Murine typhus is most common in the southeastern and Gulf Coast states and in southern California.

487. **Is epidemic typhus the same thing as murine typhus?**

    No. Epidemic typhus is also a rickettsial disease, but unlike murine typhus, epidemic typhus is transmitted by a louse and is not associated with rodent vectors or reservoirs. Although the symptoms with the two diseases are similar, epidemic typhus is more severe.

Plague

488. What is plague?

Plague is a deadly bacterial disease that kills fleas, rats and humans. The specific bacterial species is *Yersinia pestis*, formerly known as *Pasteurella pestis*.

489. Does plague have another name?

Yes. It is also called "Black Death," murine plague and urban plague.

490. How is plague transmitted to humans?

Specific species of fleas draw contaminated blood from the rodent. The bacteria thrive in the flea, and when the flea goes to feed off the human, it injects the bacteria directly into the human bloodstream. Bacteria can also enter by rubbing contaminated flea feces into cuts. One hundred million plague bacteria can accumulate in one milliliter of rat blood.

491. What domestic rodents are associated with this disease?

Roof rats and Norway rats. The house mouse is not considered an important vector in the United States, but has been implicated in Russia, Indochina and Brazil.

492. What wild (native) rodents are associated with the plague?

Plague is associated with the following feral species in the United States: deer mice, meadow voles, squirrels, chipmunks, wood rats, marmots and prairie dogs.

493. What is sylvatic plague?

Sylvatic, or campestral, plague is identical to urban plague except the animals involved in transmitting sylvatic plague are wild (native) rodents.

494. Is sylvatic plague deadly to animals?

Yes.

495. When did plague first occur in humans in the United States?

From 1900 to 1904 in San Francisco.

70 VERTEBRATE PEST HANDBOOK

496. When was the last rat-borne plague outbreak in the United States?

   From 1924 to 1925 in Los Angeles.

497. What fleas carry the plague organisms from rat to man?

   The Oriental rat flea, *Xenopsylla cheopis,* is the major vector.

498. What symptoms are prevalent in humans if they contract this disease?

   Humans can contract three different forms of plague—bubonic, pneumonic and septicemic. With bubonic plague, the lymph glands (buboes) swell severely. With pneumonic plague, the patient coughs and spreads the bacteria into the air, resulting in an epidemic. Septicemic plague affects the blood stream.

499. What type of plague is present in the United States?

   Bubonic plague is present in the western part of the United States in native rodent ground squirrels and some other wild rodent species. The plague bacillus can survive for long periods in the burrows and nests. It is transmitted by fleas from one rodent to another.

500. Do people still contract plague in the United States?

   Yes. There are a few cases yearly in the western states, when people come into contact with wild rodents. Most people recover after antibiotic treatments, but there are occasional fatalities.

   Fleas are normally the transmitting agent, although occasionally plague is acquired from handling an infected rodent carcass or from infected animals such as cats.

Rickettsialpox

501. What is rickettsialpox?

   This disease is caused by a rickettsia, *Rickettsia akari.*

502. How is rickettsialpox transmitted?

   A small mite found on the house mouse carries the rickettsia from mouse to mouse and from mouse to people.

503. **Where has rickettsialpox occurred in humans in the United States?**

The disease has occurred in Boston, Cleveland, New York, Philadelphia and West Hartford, Conn. New York leads the country in cases, with more than 100 per year.

504. **What domestic rodents are involved?**

House mice.

505. **What symptoms are prevalent in humans if they contract the disease?**

Besides a papular lesion, humans develop chills, fever and headaches, followed by a rash over most of the body.

CHAPTER 6
Non-Commensal Rodents

Cotton Rats
(Sigmodon)

506. How big are cotton rats?

    They are smaller than a Norway rat but bigger than a house mouse or meadow mouse. They weigh about four or five ounces. The cotton rat's tail is shorter than that of a Norway rat.

507. What do cotton rats feed on?

    These rodents eat numerous field crops, including melons, tomatoes, alfalfa and cotton.

508. Where are cotton rats found?

    They frequent fields and like to make nesting burrows under trash and rubble. Geographically, they are found in the United States from the Virginias south and west to Texas and into Mexico.

509. Why are these rodents called cotton rats?

    In the South where they dwell, they are known to feed on cotton plants.

510. Are cotton rats nocturnal?

    Yes, but some are active during the day.

511. Do cotton rats feed on bird eggs, such as quail?

    Yes.

Non-Commensal Rodents

512. Can cotton rats be live trapped?

    Yes, a good bait for live or snap traps is peanut butter or cat food on carrots or sweet potatoes.

513. Can cotton rats be trapped with rat-sized snap traps?

    Yes.

514. What other crops do cotton rats damage?

    They are a problem in sugar cane and melon crops.

515. What is the life cycle of the cotton rat?

    They breed throughout the year, having two to 15 young per litter. The gestation period is 27 days. At 10 to 15 days, the young are weaned. They are sexually mature at two to three months of age.

516. How long does the average cotton rat live?

    About six months.

517. Do cotton rats spread any diseases?

    Yes. They have been incriminated in the spread of hantavirus in the southern United States.

518. How are cotton rats controlled?

    Cotton rats will take most rodenticides if applied according to the label directions for commensal rodents. Only zinc phosphide products are specifically labeled for this species. Zinc phosphide baits may be applied to runways or scattered into crop fields.

Deer Mice and White-Footed Mice
(Peromyscus)

519. Are deer mice also called white-footed mice?

> No. They are sometimes mistaken for each other because of their similar appearance, but they are separate species in the genus *Peromyscus*. The white-footed mouse is *Peromyscus leucopus*. The deer mouse is *Peromyscus maniculatus*.

520. What are the distinguishing physical characteristics of mice in this genus?

> They have large dark eyes, ears larger than a house mouse's, and they usually have a white belly. Their feet end in a distinct line and have a tawny-colored upper surface. The tail of *Peromyscus* is covered with fur such that the bottom of the tail is white, and the top dark. (House mice have naked tails.)
>
> Differentiating deer mice from white-footed mice is difficult. Refer to specialized texts for diagnostic aids, or use some of the behavioral differences outlined in the following section to help make the distinction.

521. With what disease are these two species associated?

> Both species are associated with hantavirus. The deer mouse is also the primary vector of the Lyme disease organism. (See questions on both diseases in preceding sections.)

522. Are there other species of *Peromyscus*?

> Yes. At least nine species belong to the genus in North America.

523. Do deer mice commonly invade structures?

> Yes.

524. Are deer mice found in cities?

> Yes, deer mice are more common in urban and suburban habitats than people realize.

525. Do deer mice feed on gypsy moth caterpillars?

> Yes.

Non-Commensal Rodents 75

526. **Do white-footed mice invade structures?**

    White-footed mice invade structures much less frequently than deer mice. They spend most of their lives in wooded-field areas.

527. **Do *Peromyscus* mice have fur on their tails?**

    Yes, typically dark on top and white on the bottom.

528. **What is the life cycle of both the deer mouse and white-footed mouse?**

    The gestation period is 21 days. The young are weaned at 14 to 20 days. The female is mature at 28 days and can conceive at 39 days. The adult male takes 10 days longer than the female to mature.

529. **How long do *Peromyscus* mice live?**

    They can survive more than five years, but normally live less than one year. Even at about 33 months old, they can still breed.

530. **How do you control deer mice inside a structure?**

    The use of snap traps avoids odor problems. Use peanut butter and rolled oats for bait. Set traps the way you would for house mice. Glue boards and multiple-catch mouse traps also work well.

531. **How do you control deer mice outside a structure and keep them from invading buildings?**

    Keep weeds and brush away from structures, and store firewood and other potential harborage materials at least 100 feet from the building. Rodent-proof small openings. *Peromyscus* mice can be killed with most rodenticides if used according to the label instructions for house mouse baiting. Zinc phosphide is labeled for control of *Peromyscus* mice away from structures, including in crop fields.

True or False

532. Deer mice are seed feeders. — **True**

533. Deer mice have long tails. — **True**

534. Deer mice have large eyes and big ears. — **True**

535. Deer mice are active during the day. — **False**—Like most rodents, they are active principally at night (nocturnal).

536. Deer mice cut runways through the grass. — **False**—Meadow voles and some other rodents are more known for this.

537. Deer mice commonly make nests in hollow tree stumps. — **True**

538. Deer mice often enter houses in winter for shelter. — **True**

539. During the early spring in a wet year, deer mice may move their young and nest up into trees, such as into the base of squirrel nests. — **True**

540. Rodents in the genus *Peromyscus* are reservoir hosts for the hantavirus disease. — **True**

541. Rodents in the genus *Peromyscus* are associated with ticks that carry Lyme disease. — **True**

Meadow Voles and Pine Voles
(Microtus) and (Pitymys pinetorum)

542. Are meadow voles sometimes called meadow mice and field mice?

    Yes, they belong to the genus *Microtus*.

543. What do voles look like?

    They are stout and have short legs and a short tail. Their eyes and ears are small. They are about twice the size of a house mouse or deer mouse and weigh about two to three ounces. They are usually brown or gray, but many color variations exist.

544. Is there more than one species of vole?

    Yes. There are about seven different species in the United States. The home range of some species overlap. The meadow vole is one species with a wide range across the northern states.

545. Are meadow mice found in all areas of the continental United States?

    No. There are no voles in the Gulf states.

546. When are voles active?

    Voles are active during both the day and the night. Their daytime activity may be in response to the many predators that feed on voles at night, such as hawks and owls.

547. What kind of burrow system do voles create?

    Many vole species such as meadow voles create surface runways and tunnels leading to multiple burrow entrances.

548. Are voles good swimmers?

    Yes. They often live in marshy areas.

549. Do meadow voles live inside structures?

    Only rarely do meadow voles move indoors. More commonly, they inhabit fields with thick grass. They may be found tunneling under planters, boards or other objects near buildings and may rarely invade.

550. **What do meadow voles feed on?**

   Voles feed on grasses, seeds, roots, bulbs and bark. They are fond of fruit such as apples, and they will also feed on insects.

551. **What is the vole life cycle?**

   They can produce up to 17 litters per year in the lab, but in the field will generally have from one to five litters, with four to eight young per litter. The female vole has a 21-day gestation period. Young are sexually mature in 30 to 45 days.

552. **How long after giving birth can voles become pregnant again?**

   The female is ready to conceive again almost immediately after the birth of a litter. Under the best conditions of weather, habitat and food, this results in tremendous numbers of voles being produced in a very short period of time.

553. **Do vole populations fluctuate?**

   Yes. In many areas, vole populations build and crash in a two- to five-year cycle.

554. **How many voles can exist in an acre field?**

   Twelve thousand voles have been recorded in agricultural lands during a vole population high.

555. **How long do voles live in the wild?**

   About one to one-and-one-half years.

556. **Can voles become an economic pest?**

   Yes. When populations are high, sometimes, under the protection of snow cover, they can invade areas around building foundations, causing damage to bushes and landscaping. They can also damage turf and golf courses, particularly under snow cover. They can be very destructive in fruit tree orchards, especially during the fall and winter when other sources of food are limited. During these times, voles burrow around the base of trees and eat the roots and tree bark. Trees girdled by voles are killed frequently. Voles can also invade many field crops and cause damage.

Non-Commensal Rodents 79

557. **How can you distinguish tree girdling by voles from damage caused by other animals?**

    Voles leave an irregular pattern. Vole gnaw marks on the tree are at various angles and in irregular patches. The location of damage from a vole often corresponds to the maximum level of snow cover. Rabbits, deer and other foraging animals usually cause damage higher on the tree and leave larger gnaw marks of a more defined type.

558. **What signs do you look for if you suspect voles?**

    Many vole species, such as the widespread meadow vole, produce extensive runway systems just beneath the surface with many burrow openings. Runways are one to two inches wide. Active runways and burrows will have freshly excavated soil nearby. Voles are often detected after snow melts by remnants of surface trail networks and excavated dirt left on the surface.

    In warmer areas, some vole species do less burrowing and are more active above ground. The pine voles in the southern United States are an example.

559. **Is the pine vole a *Microtus* species?**

    Although it resembles other voles, pine voles have been placed by some authorities into a separate genus *Pitymys*, species *pinetorum*.

560. **Can voles be trapped?**

    Voles are difficult to trap because they tend to stay in their burrows and rarely venture about on the surface. Burrow systems usually need to be excavated in order to accommodate snap traps or small live traps. Voles can fill or cover traps rapidly with dirt and render them inoperative. A few voles near a building may be removed by trapping, but this is not practical on a large scale. Mouse snap traps should be placed perpendicular to the runway system with the trigger end in the runway. Apple slices or peanut butter and molasses or peanut butter-oatmeal mixtures make good baits. Trapping success can be improved by prebaiting, using sprung or wired-open traps.

561. How else can vole damage be reduced?

Voles can be kept from valuable plantings by the use of hardware cloth cylinders around the base or stem of plants or saplings, buried at least six inches in the ground. The mesh openings should be one-quarter-inch or less. Reducing habitat around structures or valuable plantings can limit vole numbers. Mow lawns and turf regularly, and remove mulch from the base of trees. Tilling the soil around plantings removes cover, destroys burrow systems and can even kill existing voles. Repellents, however, have not been proven very effective for long-term prevention of vole problems.

562. Can voles be controlled with rodenticides?

Yes, rodenticides have been a mainstay for large-scale control of voles in agricultural situations. Zinc phosphide products are labeled for the control of voles, and some states have labels allowing apple producers and tree farmers to broadcast or hand-place various anticoagulant baits around trees to protect them from vole damage. Voles causing damage to turf and landscaping around buildings can be controlled by rodenticides. Rodenticides can be placed in the burrow system or underneath a board or cover where voles are active. Voles can be controlled effectively with nearly all rodenticides labeled for the control of commensal rodents. Follow label directions and bait for house mice (baiting in inaccessible areas or bait stations). Pellets are typically better accepted by voles than wax blocks, although pellets may not hold up well when in contact with the soil. Use of place-packs can improve the weatherability of pellet bait placements.

Muskrats
(Ondatra zibethica)

563. Where is the muskrat found?

    Throughout most of the United States except in arid, desert areas.

564. What type of habitat do muskrats prefer?

    Muskrats are usually found near water, but are capable of making long trips over land to food sources or preferred habitat.

565. Can muskrats swim?

    Yes, muskrats are among the most aquatic of rodents. They often build their homes in river banks with an underwater entrance. Their hind toes are webbed for swimming.

566. How big is a muskrat?

    Muskrats weigh from two to three pounds and are about two feet long. Sightings of muskrats often account for tales about "rats as big as cats."

567. How long do muskrats live?

    They live from four to five years.

568. What is the life cycle of the muskrat?

    Adults breed when they are from one to one-and-one-half months old. The gestation period is about 29 days. There are four to six litters per year and usually two to six young per litter, but they may have as many as 15 in a litter.

569. What do muskrat droppings look like?

    They are elongated and about five-eighths of an inch long.

570. What do muskrats eat?

    They feed on roots, stems, aquatic leaves, insects, worms and shellfish. They also eat fruits and vegetables and may consume dead fish or carrion.

571. Do muskrats make a nest?

    Yes, they construct a nest of sod, twigs and mud.

572. When are muskrats most active?

   They are most active at dusk and early morning during the spring.

573. Are muskrats as agile and active as Norway rats?

   No, their movements are more slow and sluggish.

574. Are muskrats valuable?

   They are often trapped for their fur. Prices for the fur can vary greatly from year to year, and generally they are not as valuable today as they were in the past, due to changes in fashion. Areas where trapping has declined may see muskrat population surges that can lead to damage.

575. What economic damage do muskrats cause?

   They may invade crops, and their burrowing may damage the banks of ponds or levees.

576. How can muskrats be controlled, such as at a farm pond?

   Drop the level of the water in the pond by about two feet during the winter to expose the burrows. Fill the burrows, dens and runs with stones, and set traps. No. 1 steel traps or Conibear traps are effective. Consult your local extension office or Fish and Wildlife Service for trapping regulations and tips effective in your area.

577. What toxicants are available for the control of muskrats?

   Zinc phosphide products are the only ones registered for control of muskrats. This is a restricted-use pesticide. A zinc phosphide concentrate can be applied with a vegetable oil sticker to cubes of apples, sweet potatoes or carrots. Baits are typically placed on floating platforms in burrow entrances or near feeding areas.
   Some states have permits allowing the use of commercially available anticoagulant baits. Some success has been achieved using wax block products fastened to floating rafts.

Nutria
(Myocastor coypus)

578. What is a nutria?

It is a rodent from South America belonging to the family *Capromyidae*. The Latin name is *Myocastor coypus*.

579. Is it called by another name?

Yes, in Brazil it is called the Coypu.

580. Is it present in the United States?

Yes, it is now established in the wild from Texas to Florida and north to Tennessee. It is also found in parts of Oregon, Washington, Idaho, Utah, Colorado, New Mexico and Nebraska. On the East Coast, it is in North Carolina, Virginia and Delaware.

581. How did this rodent get to the United States?

It was brought in and bred in captivity for its fur. Some animals escaped, bred and spread throughout the country's warmer areas.

582. Are nutria an economic pest?

Yes, especially in the Gulf Coast area, where they feed on sugar cane and rice and cause thousands of dollars of yearly damage.

583. Do nutria have value?

There is a considerable industry in the Gulf Coast states selling nutria for their fur, and their meat is used for processing into pet food.

584. What is the size of nutria?

These rodents weigh from seven to 12 pounds, and some individuals can weigh up to 20 pounds. In size, they rank between muskrats and beavers.

585. How are nutria controlled?

Nutria can be trapped. In some areas, they are also poisoned with zinc phosphide on carrots. Check on local laws with your extension service or Fish and Wildlife office.

586. **If I use rodenticides for nutria, how is the bait placed?**

   Commonly, bait is placed on floating rafts spaced about one-quarter to one-half mile apart throughout the area. Up to 10 pounds of bait is placed on each raft.

587. **Where can I obtain more information on the biology and control of this animal?**

   In addition to the general sources given at the end of this book, two good sources are:

   "Nutria Control Using Zinc Phosphide," *Pest Control* magazine, September 1971, pp. 30, 32, 34, 36, 37.

   "About Nutria and Their Control," by James Evans. For copies, write to: Denver Wildlife Research Center, Denver, Colo. 80225.

Pocket Gophers
(Geomys and Thomomys)

True or False

588. Pocket gophers tunnel in the ground. — **True**

589. In the North, pocket gophers can tunnel under snow. — **True**

590. Gophers form chains of dirt on the surface of the ground. — **True**

591. Pocket gophers live in colonies in a burrow system. — **False**—Other than at mating time, they live singly with one animal per burrow system.

592. Pocket gophers have a home range of up to 700 square yards. — **True**

593. Pocket gophers hibernate. — **False**

Answer the Following

594. What is the life cycle of the pocket gopher?

   Pocket gophers reach sexual maturity in the spring. In the South, they can have two litters per year, and in the North one litter. Litter sizes range from one to 10 pups, averaging three to four. The gestation period is 18 to 19 days, but up to 51 days for the Plains pocket gopher.

595. What is a salamander?

   Normally it refers to an amphibian. However, in some parts of the South (especially in Florida), it is a nickname for the pocket gopher. Land tortoises in the Southeast are also incorrectly called "gophers."

596. What is a "striped gopher"?

    This rodent is not a gopher at all, but instead is the 13-lined ground squirrel, *Citellus tridecemlineatus*.

597. What do pocket gophers eat?

    They are vegetarians. In nature, they feed on tree roots below the surface of the earth.

598. Why are they called pocket gophers?

    They have fur-lined cheek pouches in which they carry food.

599. Can you drown gophers by filling their burrows with water?

    No. This will drive gophers from their runways, but few will drown.

600. Can gophers swim?

    Yes, they are good swimmers.

601. Are there any plants that repel gophers in order to keep them out of a garden?

    There are some plants sold for this purpose. They give mixed results. The best known is called gopher's spurge, *Euphorbia lathyrus*. It is sold under a variety of names, including Nature's Farewell and Gopher Patrol. The roots are poisonous to gophers and it is necessary to plant a solid wall of plants for an effective barrier. Castor-oil plants are also considered repellent. These plants may also deter moles.

602. Can you trap pocket gophers?

    Yes, but it is a complex process. There are various traps on the market with varying degrees of effectiveness. The guillotine or spear type of trap is reasonably effective. Spring traps that must be placed in gopher burrows give mixed results because gophers can be reluctant to re-enter the disturbed areas.

603. Can toxicants be used to control pocket gophers?

    Yes. Currently, there are strychnine and zinc phosphide baits labeled for this use. Some states allow the use of anticoagulant baits as well. Baiting for gophers is considered less hazardous than some other applications because all bait is placed below ground. For treating large acreage, there are special bait-applying machines that can be pulled behind tractors, creating an artificial burrow and dispensing bait into the burrow as it goes.

There are also some fumigant cartridges available that are lighted in gopher burrows to release toxic gas in the burrow system. These are only partially effective, because gophers can plug up parts of the burrow network and prevent the gas from entering.

Porcupines
(Erethizon)

604. Why are porcupines so named?

    In Latin, *porcus* means "swine" and in French, *epine* means "thorn."

605. Do porcupines "shoot" their quills?

    No. When attacked, they slap their tail and the loosely attached quills are embedded into the person or other animal.

606. How big do porcupines get?

    They weigh up to about 25 pounds.

607. Is there more than one species of porcupine in the United States?

    Yes. The Canadian porcupine *Erethizon dorsatum* inhabits the Northeast as far south as Pennsylvania and westward and northward to Alaska. The yellow-haired porcupine, *Erethizon epixanthum,* dwells from the Great Plains westward to the West Coast, and from Alaska south in the Rocky Mountains to southern Arizona.

608. How many quills are there on an adult porcupine?

    They possess about 30,000 quills.

609. Can porcupines climb trees?

    Yes, they are excellent climbers and climb trees to escape and sleep. They are also very destructive because they eat tree leaves and twigs.

610. Do porcupines have natural enemies?

    Mountain lions and bobcats will climb trees and kill porcupines. Dogs, wolves and coyotes will attack porcupines, but are often bested.

611. When are porcupines most active?

    They are most active at night (nocturnal). During the day, they commonly rest in trees or caves.

Non-Commensal Rodents 89

612. Do porcupines hibernate in the winter?

    No.

613. Should porcupines be killed when found in an area where people live?

    Unless the animals are causing a problem, they should be left alone. They never attack people. Consult your local conservation official before undertaking a control program.

614. How are porcupines normally controlled?

    Common techniques include trapping and shooting. Pest-proofing doors and windows, as well as erecting fencing, can keep them out.

615. Can porcupines be caught in live traps?

    Yes. Good baits include apples or carrots. Porcupines are also strongly attracted to salt, so traps can be baited with simply a salt-soaked cloth, sponge or piece of wood.

616. Once caught, how far away should a porcupine be released?

    In states where trapping, transport and release of porcupines are legal, at least 25 miles. Generally, the transfer of wildlife to new areas is not recommended because of the risk of spreading animal diseases, and the stress on the animal from having to adapt to a new home.

617. Are there any registered toxicants or repellents for porcupines?

    No, but repellents sold for deer, squirrels or rabbits may have some effect on porcupines.

Prairie Dogs
(Cynomys)

618. Are prairie dogs rodents?

    Yes.

619. Is there more than one species of prairie dog?

    Yes. There are several species, including the black-tailed prairie dog and the white-tailed prairie dog.

620. Why are they called prairie dogs?

    They live on the prairies of the Western Plains region of the United States. They produce a sound like a bark or an alarm cry. (They can also produce a high-pitched smooth note.)

621. Do prairie dogs live in colonies?

    Yes, and in some cases they are government-protected, as in the national parks in North Dakota, South Dakota, Wyoming and Oklahoma.

622. Are prairie dogs ever controlled?

    Yes. On some occasions, prairie dogs can become a nuisance on private property.

623. When necessary, how are prairie dogs controlled?

    Because they are protected animals, it is best to check with the National Fish and Wildlife Service before undertaking a control program that involves baiting, trapping or fumigation.

624. Are there special concerns about controlling prairie dogs?

    Yes, because many other animals use the prairie dog burrow systems. One such animal is the endangered black-footed ferret. This ferret is strictly protected, and no control operations can be conducted anywhere that they are found.

Squirrels and Chipmunks
(Sciuridae)

625. Is there more than one species of tree squirrel?

Yes. There are four types of tree squirrels in the eastern United States and several subspecies within each type.

Gray squirrels, fox squirrels, red squirrels and flying squirrels are all tree squirrels. Here are their Latin names:

Gray squirrels—*Sciurus carolinensis*
Fox squirrels—*Sciurus niger*
Red squirrels—*Tamiasciurus hudsonicus*
Flying squirrels—*Glaucomys volans* and *G. sabrinus*

In the far western states (Washington, Oregon and California) western gray squirrels (*Sciurus griseus*) are prevalent. There is also an Apache fox squirrel, *Sciurus apache,* in northern Mexico. Two species of squirrels with long, tassel-like ears inhabit the Grand Canyon area. These are the Kaibab and Abert squirrels.

626. What is the life cycle of a tree squirrel?

Squirrels can breed throughout the year, but in the North there is a definite correlation between breeding and warmer weather. There are one to six offspring per litter. The adult female carries her young for 40 to 45 days. The young squirrels remain in the nest for about six weeks. There is one litter per year in the early spring in the North, but two litters are common in the warmer states.

627. What is the favorite food of gray squirrels?

Acorns. They also feed on other nuts, seeds, berries, bulbs, leaves, insects, bird eggs, small birds and mammals. They are attracted to backyard bird feeders, corn and dry dog and cat food.

628. How heavy does a gray squirrel get?

They can weigh about one pound.

629. How good a jumper is the gray squirrel?

They can jump eight feet sideways, four feet straight up and 15 feet downward on a tree trunk.

630. What is a black squirrel?

A color variation of the fox or gray squirrel.

631. What is a flying squirrel?

Flying squirrels refer to a group of squirrels who possess a broad fold of furred skin that extends on both sides of the body when the squirrel glides through the air. The tail hairs extend laterally to help in gliding. Flying squirrels belong to the genus *Glaucomys*.

632. What is a ground squirrel?

Ground squirrels, as the name implies, nest and live on the ground. They belong to the genus *Citellus* (*Spermophilus* in some texts). These squirrels are found in the West, and as far east as Ohio. In the Midwest, the 13-lined ground squirrel is quite common. Most ground squirrels make a whistle-like noise.

633. What color are fox squirrels?

They are reddish in color.

634. Can fox squirrels cause damage?

Yes, they can strip the bark off trees.

635. Do tree squirrels do any damage to property?

These rodents destroy garden bulbs and hardwood plantings by chewing on them. They will also eat fruits on fruit trees. They can kill or deform ornamental shrubs. Gnawing on the side of a building is not uncommon for them, and they can damage wiring. Once inside a structure, squirrels chew on wires and tear up papers for nesting material. If you try to catch a frightened squirrel in a room, it can result in torn furniture, broken lamps and other damage.

636. Do squirrels and chipmunks carry any parasites that might bite humans?

Yes, particularly fleas.

637. Can a squirrel or a chipmunk bite you?

Yes. They are rodents and when they bite, it is painful. One should not attempt to touch a wild squirrel.

638. Can squirrels transmit rabies to humans?

There are no cases of rabies being transmitted by rodents in the United States. However, you should consult your doctor if you are bitten.

639. Why do squirrels enter a structure?

For either food, shelter or both.

640. If present in a home, when would you be most likely to hear them?

Tree squirrels are diurnal. That is, they are active only during the day. During the night, they sleep. However, if the squirrel becomes trapped inside a wall, it will make noise until it escapes or dies.

641. Are there any repellents to drive squirrels and chipmunks out of a structure?

Once inside, very little repels them. Occasionally, bright lights or a loud radio will move them out.

642. Is it dangerous to use mothballs in an attic to drive squirrels out?

Yes. The fumes from the paradichlorobenzene are hazardous to humans. Mothballs are not labeled for this use. The vapors are heavier than air and can invade the lower living spaces.

643. What should be done once squirrels or chipmunks are driven out?

It is important to close up any entrance holes before they return. Use sheet metal, steel wool or wire mesh (screening).

644. If squirrels are chewing on wood or plastic garbage cans, is there any way to stop them?

Yes. There are repellent products that can be applied to the wood or garbage can. One is RoPel®, containing denatonium saccharide. It is also possible to make a paste containing tar to apply to wood or other materials, and that is usually effective as a repellent. The repellent is made as follows: Three pounds of asphalt emulsion, two quarts of water and two pounds of copper carbonate. Mix the water with the asphalt, then add the copper carbonate. Apply with a brush. The mixture is black and so may not be suitable for all surfaces. There are also sticky caulking products, such as J.T. Eaton's 4 The Squirrel™ repellent (polybutene), that can be used on wires and window sills.

645. **Are there repellents for keeping squirrels and chipmunks off plants?**

There are some repellents that can be sprayed on plants and fruit trees to keep deer, squirrels and other animals away. One is Hinder®, containing ingredients similar to rotten eggs.

646. **Is there a repellent to keep squirrels from eating birdseed?**

Yes, there are several products available at nurseries and garden stores that can be mixed with birdseed that reduce squirrel consumption. One is called Squirrel Away™, made by Scrypton Systems, Inc. Most of these products contain capsicum, the active ingredient in hot peppers. Mammals find this hot material distasteful, but birds are not bothered by it.

647. **Is there a way of building squirrels out?**

Yes, but it is necessary to use one-half-inch mesh or 26-gauge metal or heavier to screen them out. Look for eave openings, unscreened attic vents, knotholes, loose flashing around chimneys and vent pipes, and openings around cables.

648. **Are there any special techniques in live trapping squirrels?**

Squirrels are not always easy to live trap, especially inside a structure. It is sometimes necessary to leave the trap wired open and baited for several days before the squirrel will go inside, and then rebait and set the trap when the bait has been taken. Placing a trail of peanuts or other attractive bait in front of and into the trap will sometimes help encourage the squirrel to enter. One pest control operator, it is reported, used a red rubber ball to trap flying squirrels.

649. **How can you best keep the squirrel inside the trap once caught?**

Squirrels can flip the traps over, thereby releasing the doors in some models. Use a metal bar running perpendicular to the length of the cage. Run the bar through the cage so that the ends stick out both sides at least a foot. This will stabilize the cage so that it cannot be flipped over. Placing a weight on the trap door of some models will drop the door more quickly.

650. **If no holes are found but squirrels are inside the attic, where should you look for entrances?**

    The facia boards behind the rain gutters (particularly behind the downspouts) may rot out from ice buildup, creating an opening.

651. **When setting a live trap for squirrels living in an attic, is it better to set the trap inside the attic or on the exterior?**

    Exterior trapping has a few advantages. It makes it easier to check the trap without the homeowner being present. It can also sometimes be checked from the ground, and squirrels enter baited traps more readily outside than indoors.

652. **What is a good bait for live trapping squirrels?**

    Besides acorns, nut meats work well. Sunflower seeds, unroasted peanuts, and peanut butter and molasses on bread will attract them.

653. **Are there any traps available that can be used to kill squirrels?**

    Squirrels are protected game animals in many states and municipalities, and it may be illegal to kill them. Check with your local extension service or wildlife official. If they can be killed, then No. 0 or No. 1 steel traps or Conibear traps may be used.

654. **What happens if a squirrel dies in a wall?**

    The carcass will decompose and begin to smell in a few days. Blow flies will invade the carcass if they can reach it. After a week or so, maggots will migrate away from the carcass and are sometimes seen crawling on the floor. Keep in mind that squirrels may also become trapped and die in chimneys.

655. **Once a squirrel is live trapped, how far do you have to take it from the site to ensure that it will not come back?**

    Squirrels have been known to return from several miles away. It is best to take them 15 or 20 miles from the place of capture for release. Check with state regulators to determine whether the relocation of squirrels is permissible in your area.

656. How do you keep a squirrel from going up a tree whose branches overhang a building?

Place a two-foot-high band completely around the tree so that the bottom of the band is six feet off the ground. Also clear a distance of six feet between the roof of the structure and overhanging branches. Make the band adjustable and loosen as the tree grows.

657. Can squirrels walk over telephone wires to reach a structure?

Yes, they are very adept at walking on wires.

658. Are squirrels and chipmunks protected?

Some states and local areas protect these animals, thereby allowing only live trapping. Consult your local regulations before undertaking any squirrel control program.

659. Can you catch squirrels on glue boards?

This approach sometimes works where it can be done legally. Glue boards should be fastened down or tied to wire so the squirrel does not crawl away.

Chipmunks
(Tamias)

660. **Are there different types of chipmunks in the United States?**

    Yes. The Eastern chipmunk, *Tamias striatus*, is found in every state east of the 102nd Meridian, except Texas and Florida. In the West, different species of chipmunks are present, once given a different genus *Eutamias*. All chipmunks are now considered to be of the *Tamias* genus. The biology of all the chipmunk species is very similar.

661. **How many chipmunks have been known to enter a single home?**

    The most known to this author is 85 chipmunks, all live trapped out of and around one home.

662. **Are chipmunks solitary animals?**

    Only in the fall, when they fight to defend their territory.

663. **How do chipmunks' nesting habits differ from those of tree squirrels?**

    Chipmunks nest underground in a burrow system that may extend for 30 feet or more. Tree squirrels nest in trees, hollow logs or buildings.

664. **Where is a good place to look for a chipmunk nest?**

    Burrow openings are often located at the base of stumps, rocks, walls, logs or shrubs.

665. **What is the life cycle of chipmunks?**

    They have two to six young per litter. Females have a gestation period of about 33 days. The young remain in the underground nest for four to six weeks.

666. **What do chipmunks feed on?**

    They prefer to eat seeds and berries, but will also eat green vegetation.

667. **In live trapping chipmunks, what is a good bait to use?**

    Prune pits, unroasted peanuts, corn, sunflower seeds, peanut butter, cereals, grain and granola bars all will work.

668. Do anticoagulant rodenticides kill chipmunks?

There are no rodenticides labeled for the control of nuisance chipmunks.

True or False

669. Adult chipmunks weigh two to four ounces.

    True

670. Norway rats and chipmunks both have four toes on their front feet and five toes on their hind feet.

    True—It can be hard to distinguish rat tracks and chipmunk tracks.

671. Chipmunks mate twice a year.

    True

672. Adult chipmunks live up to three years.

    True

673. Chipmunks are active during the day.

    True

674. Chipmunk burrows are about two inches in diameter.

    True

675. Snap traps can catch chipmunks.

    True—In areas where this can be done legally. Traps should be covered with a box so non-target animals are not caught.

676. Chipmunks can do extensive damage to a structure.

    True—They are very aggressive burrowers and can undermine stone foundations or driveways. There have been instances of a number of chipmunks doing more than $50,000 worth of damage.

677. There are special traps for chipmunks. **True**—The Kness Manufacturing Co. came out with such a trap in 1993. Other manufacturers have live traps that can be used for different rodents about the size of a chipmunk.

678. If you live trap a chipmunk for release, you must take it several miles away or it will return to its original location. **True**

679. Chipmunks may survive a snap trap hit. **True**

680. Chipmunks are relatively easy to live trap compared to squirrels, especially if prebaiting is done. **True**

681. Chipmunks are active burrowers and, as such, can do tremendous damage to a driveway, slab or retaining wall. **True**

682. When digging, chipmunks carry the dirt away in their cheeks and spread it out, thereby avoiding early detection. **True**

683. There are no poisons registered for chipmunk control. **True**

Woodchucks
(Marmota monax)

684. Is there another name for the woodchuck?

    Yes, in some regions it is called a groundhog, whistle pig or lowland marmot.

685. Where is it found?

    In the United States, the woodchuck is found in the Northeast and Midwest. Its range goes westward to Idaho and Kansas, and south into Alabama and Virginia.

686. Is there a comparable species in the western states?

    Yes, in the West there is the yellow-bellied marmot, *Marmota flaviventris*. These marmots live on high mountains among the rock slides and cliffs. There are also species (olympic and Vancouver marmots) found in the extreme Northwest (state of Washington) that range up into Canada and Alaska.

687. How heavy are woodchucks?

    Woodchucks can weigh up to about 15 pounds or more.

688. How long do woodchucks live?

    They can live about nine years.

689. How old are woodchucks before they can breed?

    About one year old.

690. How long do they carry their young before birth?

    Twenty-eight days.

691. How many young do woodchucks have per litter?

    There are three to five young per litter.

692. Do woodchucks build their own dens by burrowing?

    Yes.

693. Where are woodchucks commonly found?

    In open fields, pastures, meadows and woodlands.

694. **Where do woodchucks rest?**

    In their burrows or in hollow logs.

695. **Do woodchucks come out during the day?**

    Yes, they are diurnal.

696. **Can woodchucks feign death?**

    Yes.

697. **Are woodchucks good burrowers?**

    They are excellent burrowers. Their burrows may be six feet deep and may extend for 25 feet or more.

698. **What kinds of damage can woodchucks do?**

    In agricultural land, their burrows can damage equipment and injure livestock. They can also consume crops and plantings.

699. **Can woodchucks climb trees?**

    Yes.

700. **What do woodchucks eat?**

    They prefer various grasses, clover and succulent green plants.

701. **How are woodchucks controlled?**

    In areas where they do damage or constitute a nuisance, they are generally killed by trapping or fumigation. Woodchucks are not protected animals in most of the United States.

702. **What types of traps are used to kill woodchucks?**

    No. 1-1/2 or No. 2 steel traps are generally used.

703. **How are the traps placed?**

    Traps are placed on or near the mounds of dirt leading to the burrow. Use two traps, because the rodent may spring one trap inadvertently by throwing dirt on it.

704. **Is there another way to eliminate woodchucks aside from trapping?**

    Yes. Gas cartridges are commercially available, such as the Giant Destroyer. These burning cartridges contain sulfur and other materials, and give off poisonous gases in the burrow.

Some states classify gas cartridges as fumigants and regulate their use.

705. How are gas cartridges used?

Before using, all the openings to the burrow are blocked except for one. Quickly place the ignited cartridge into the burrow opening and rapidly close the opening with dirt.

706. Why is it best to control woodchucks in the early spring?

In early spring, it is most likely that only the woodchuck is using its freshly dug burrow. As the season progresses, other animals may also use the burrow, and the risk of killing non-target animals increases.

707. Can woodchucks be live trapped?

Yes. However, because of their strength, be sure the trap is metal and of robust construction. Bait with a head of lettuce, apples, carrots, fresh beans, sweet corn or peas. If observations have indicated the path normally taken by the animal, you can improve your chances by placing rocks, logs or other objects to "funnel" the woodchuck to the trap placement, and the attractiveness of the bait will be less important. Also, trap success may be increased by lining the bottom of the live trap with fresh grass or burying it slightly in the ground.

708. Are there any toxic baits registered for controlling woodchucks?

No.

Wiring blocks in drains; burrows (refer to question 1181)

Vitamin K antidote (refer to question 1256)

Zeneca Weatherblok® XT rodenticide

Ketch-All® (refer to question 1106)

Mouse on glue board (refer to question 1124)

Bait station (refer to question 1144)

Snap trap (refer to question 1066)

Typical live trap (refer to question 1103)

Thirteen-lined ground squirrel (refer to question 632)

Black rat snake (refer to question 972)

Another way of looking at rodent control (refer to question 1034)

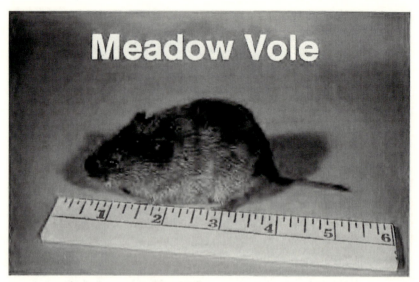

Meadow vole (refer to question 542)

Pocket gopher (refer to question 588)

Services Available From Wildlife Control Consultant, LLC

Training

We are available to speak/train on a variety of wildlife control topics. We have presented at NWCOA, Urban Pest Management Conference, Alberta Pest Management Association, Connecticut NWCO Association and others.

Writing & Research

Need help writing an article or book? Have the professionals at Wildlife Control Consultant, LLC put their experience to work for you.

Consultation

Need an outsider's opinion on a difficult topic or question?

Wildlife Control Consultant, LLC has the knowledge to help you make an informed decision.

https://wildlifecontrolconsultant.com
wildlifecontrolconsultant@gmail.com

Roof rat (refer to question 200)

House mice (refer to question 241)

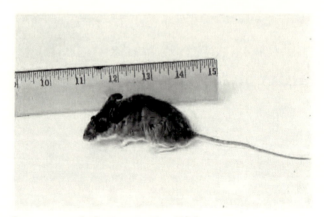

Deer mouse (refer to question 519)

Poor vision, photo courtesy of the National Pest Control Association (refer to question 162)

Rat burrow (refer to question 191)

Norway rat (refer to question 80)

Rat contamination (refer to question 100)

Rat teeth (refer to question 48)

CHAPTER 7
Pest Vertebrates Other Than Rodents

Armadillos
(Dasypodae)

709. Are armadillos rodents?

   No. They belong to a group called *Edentata*, which includes sloths and anteaters.

710. What species is in the United States?

   The nine-banded armadillo, *Dasypus novemcinctus*.

711. What is the range of the armadillo?

   This animal is extending its range northward from Mexico. It is now found in Texas, across the lower half of the Gulf states and into Florida.

712. What type of habitat does it prefer?

   It prefers to stay among rocks where there are vegetation and caves available.

713. What do armadillos eat?

   About 85 percent of their diet is insects (ants and beetles). They also eat roots and other vegetable matter.

714. What do armadillo fecal droppings look like?

   They look like small clay marbles.

715. How big do armadillos get?

In the United States, adults reach 28 inches in length plus a one-foot tail. They can weigh up to 15 pounds. Some South American species are much larger.

716. What do the newborn look like?

They have soft flexible skins. Their eyes are open and their bodies are well-formed at birth.

717. How many offspring are there?

There are always four identical young formed from the division of a single egg. Litters are either all female or all male. The young follow the mother like young pigs do.

718. Will armadillos attack humans?

No. They are very timid, and only have small teeth. When alarmed, they roll up into a ball and protect themselves with their hard suit of armor. The only vulnerable spot is well-hidden by the armored plates.

719. Is the armadillo called by other names?

Yes. Peludos, Cadassous and Apars are nicknames for armadillos.

720. Are armadillos eaten by humans?

Yes. In South America, they are considered a delicacy.

721. What types of damage can armadillos do?

They annoy people by tearing up gardens and digging holes in lawns and golf courses.

722. When are they most active?

They are most active at night. They can make 100 holes in a lawn in one night.

723. Do they have a good sense of smell?

Yes. They are particularly keen at detecting vibrations in the ground.

724. Do they run fast?

   For short distances, they can run quickly, but when alarmed they roll up into a ball.

725. Does the armadillo have natural enemies?

   Yes. Dogs and other predators will attack them.

726. Will lights or sounds repel armadillos from an area?

   No.

727. Can armadillos be trapped?

   Generally, trapping is not very successful. Steel traps used for raccoons may sometimes work, but armadillos are difficult to hold in a trap. Recommended baits include mealworms, sardines or fish.

728. What is the best way to control armadillos?

   Chemical treatment of the soil will eliminate their insect food supply. Sometimes a hunt is necessary to eliminate them from high-value areas such as golf courses.

729. Are there any precautions in handling armadillos?

   Yes. They sometimes have a form of leprosy, or Hansen's disease. They should be handled with gloves and the hands washed afterward.

Bats
(Chiroptera)

730. **Are bats really pests?**

    Although bats can be a pest problem within a structure, it is imperative to keep in mind that, in some cases, these mammals are on the endangered species list. The bat used to be targeted for destruction. Today, we look upon the bat as a valuable resource in controlling insects, and as one of nature's most interesting mammals. Where once bats were feared, today they are a tourist attraction, such as in Austin, Texas, where approximately 750,000 Mexican free-tailed bats live under a bridge. There are currently no poisons registered for bat control. Most authorities recommend pest-proofing to keep bats out of a structure, and the killing of bats should be a last resort.

731. **Are bats beneficial?**

    Yes. They feed on insects. Their manure has been processed as fertilizer. In addition, some bats pollinate or disperse flowers.

732. **Are bats harmful?**

    Sometimes. When bats invade a structure, they bring along the problems of bat manure, parasites and the possibility of rabies transmission. It is, however, a gross misconception that most bats are rabid. If a bat is carrying the rabies virus, it will eventually die from the infection.

733. **Why do bats enter a structure?**

    They invade structures for shelter.

734. **How far do bats travel to hibernate for the winter?**

    They may go up to 300 miles to find a specific cave or abandoned mine. Some bats may only have to fly a few miles. In some cases, a few bats may remain in an attic all year round. Non-hibernating species may have to fly from Canada to Mexico.

735. **Once a hibernation spot is found, will all the bats of a colony use the same structure consistently?**

    Often they will not. Part of the colony may roost other places, and not all migrate back to the structure the same year.

Pest Vertebrates Other Than Rodents 107

736. Will bats commonly hide in church steeples and belfries?

    No. These areas are usually too cold and drafty to serve as a bat nursery. Bats are more likely to harbor in the warmer, stagnant-air areas of a structure such as in the attic.

737. Does the presence of bat manure attract other bats?

    Yes, the odor is quite strong.

738. What do bats eat?

    Most bats feed on flying insects. Some bats live on fruit, others on blood.

739. How do bats catch their insect food?

    They reach out and grab the insect on the wing with their feet. They then bite it with their sharp teeth.

740. What do bat droppings look like?

    The droppings glisten from the insect fragments they contain. When squeezed, they crush easily, as opposed to rodent droppings that remain intact.

741. What is an accumulation of bat droppings called?

    Guano. In some caves and roosts, the bat guano has been removed and processed into a valuable fertilizer.

742. Do some people eat bats?

    Yes. Some Chamooroo tribe members in Guam and other South Pacific Islands eat forest bats. This practice is causing that species to become endangered.

743. How much does the smallest bat weigh?

    The tiny bumblebee bat of Thailand weighs less than an American penny.

744. How big is the largest bat?

    The largest bat, the flying fox of Java, can have a wingspan of 60 inches and weigh up to four pounds.

745. **What is the world's largest known bat colony?**

Until recently, about 30 million Mexican free-tailed bats lived in Eagle Creek Cave in Arizona. This colony has dwindled by more than 99 percent in a six-year period.

746. **How many people have died from bat-contracted rabies?**

Figures from the Centers for Disease Control (CDC) in Atlanta, Ga., report only 18 deaths in all of North America over a 46-year period. The CDC reported that four people died from bat rabies in 1997. The deaths occurred in Montana, New Jersey, Texas and Washington. Two of the people died after failing to seek medical attention that could have saved their lives. People should seek treatment as soon as possible after being bitten or coming in contact with a bat or any potentially rabid animal.

747. **What percentage of bats have rabies?**

Less than four percent of bats submitted for testing in New York state proved to be rabid, as determined by records that span from 1965 to 1989.

748. **What should someone do if bitten by a bat?**

The New York State Department of Health has the following recommendations: If a person is bitten or scratched by a bat, or there is any suspicion that bat saliva or nervous tissue has contaminated an open wound or mucous membrane, the affected area should be washed thoroughly with soap and water, and immediate medical attention sought.

An attempt should be made to capture the bat without damaging the head (which is needed to determine whether rabies is present). If the bat can be promptly tested at a health department lab, it is possible that the person bitten may not have to undergo the lengthy and painful series of vaccinations.

If the bat cannot be captured or if the test is positive, prompt treatment is indicated. This usually consists of an initial dose of human rabies-immune globulin, a dose of rabies vaccination, and additional doses at three, seven, 14 and 28 days later.

749. **Are there vaccines that can be taken before anticipated exposure to bats that will prevent acquiring the disease?**

    Yes. There is a three-dose human diploid vaccine administered in the arm over a month. Such treatment is not given routinely, but has proven useful for people who expect to be exposed to, or who handle bats on a regular basis.

750. **How large an opening can a bat get through?**

    Common bats in the United States are from three to five inches long with a wingspan of eight to 15 inches. The smaller bat species can squeeze through a crevice as thin as a pencil, or a hole three-eighths of an inch in diameter.

751. **When are bats more active?**

    At dusk, when nocturnal insects become active. There may be a second activity peak at dawn when insects are again active.

752. **Are other sounds mistaken for the sounds of bats?**

    Yes, the sound of a tree rubbing against the side of a building, or birds or even squirrels in an attic can be mistaken for bats.

753. **How can you tell where bats are roosting?**

    Look for the bat droppings on the floor. The bats are resting directly above that area.

754. **Within a structure, where might bats be found?**

    In the attic on rafters, between floors, inside window frames and near the chimney are the most common areas.

755. **What is the life cycle of the bat?**

    Most bats have one young at a time. Some have two or three. The gestation period is two to eight months. Baby bats are breast-fed until they are ready to catch their own food.

756. **Are bats' bodies covered with fur?**

    Yes.

757. **Do bats make a nest?**

    No, they carry their young around with them.

758. Are there really vampire bats?

Yes, but they are found in Central America and Mexico and not in the United States, except possibly in southern Texas.

759. What kind of damage do vampire bats do?

They attack livestock and feed on their blood. They can weaken and even kill livestock from loss of blood.

760. Do bats really fly into your hair and get stuck?

Bats will not intentionally get stuck in a person's hair. Their feet are shaped such that they might be caught inadvertently. The author has spent many hours in rooms with bats flying about and never had a bat land in his hair. Still, bats may fly around your head fairly close outdoors. They are probably attracted to the mosquitoes attracted to you, and are not interested in you directly.

761. Will bats attack humans?

Normally, no. However, any animal with rabies, including a rabid bat, may act in an unusual and aggressive fashion. Also, if a bat with rabies starts to be affected, it may become unable to fly. If a person picks up such a bat that is still alive, he or she may be bitten.

762. Where do bats hibernate?

They hibernate in caves below the frost line and in hollow trees. Sometimes the big brown bat hibernates in walls, often undetected.

763. Are bat populations increasing?

No. Bat populations appear to be decreasing. Air pollutants from factories are believed to be a major factor in reducing natural bat populations. Humans destroying bats can also depress populations.

764. Are bats rodents?

No. Bats are mammals belonging to the family *Chiroptera*.

765. How far can a bat fly?

Some species can migrate long distances in the fall.

766. Why do some bats fly around lights at night?

   The bats are catching insects that are attracted to the lights.

767. Do bats have scent glands?

   Yes, many bat species have dermal scent glands on their heads. In some species, the odor given off is very strong and skunk-like.

768. What is the distribution of bats in the United States?

   There are about 40 different species of bats in the United States. Here is a list with scientific names and distribution:

Common Name	Species Name	Distribution
Allen bat	*Idionycteris phyllotis*	Arizona
Arizona myotis	*Myotis occultus*	Southwestern U.S.
Big brown bat	*Eptesicus fuscus*	All of the U.S.
Big free-tail bat	*Tadarida molossa*	Texas
California myotis	*Myotis californicus*	Western U.S.
Cave myotis	*Myotus velifer*	Southwestern U.S.
Eastern pipistrel	*Pipistrellus subflavus*	Eastern U.S.
Eastern big-ear bat	*Corynorhinus macrotis*	Southeastern U.S.
Evening bat	*Nycticeius humeralis*	Eastern U.S.
Florida free-tail bat	*Tadarida cynocephala*	Gulf Coast
Florida mastiff bat	*Eumops glaucinus*	Florida
Fringe-tail myotis	*Myotis thysanodes*	Western U.S.
Gray myotis	*Myotis grisescens*	Northeastern and Central U.S.
Hoary bat	*Lasiurus cinereus*	All of the U.S.
Hognose bat	*Choeronycteris mexicanus*	Arizona and California
Indiana myotis	*Myotis sodalis*	Northeastern and Midwestern U.S.
Keen myotis	*Myotus keeni*	Northeastern U.S. and Northwest Coast
Leafchin bat	*Mormoops megalophylla*	Extreme Southeastern U.S.
Leafnose bat	*Macrotus californicus*	Arizona, Nevada and California
Little brown myotis	*Myotis lucifugus*	Most of the U.S.
Long-ear myotis	*Myotis evotis*	Western U.S.
Long-leg myotis	*Myotis velifer*	Southwestern U.S.

(continued)

112 VERTEBRATE PEST HANDBOOK

Common Name	Species Name	Distribution
Longnose bat	*Leptonycteris nivalis*	Arizona and Texas
Mexican free-tail bat	*Tadarida mexicana*	Western U.S.
Mississippi myotis	*Myotis austroriparius*	Mississippi Valley and Southeastern U.S.
Northern long-ear bat	*Myotis septentrionalis*	Eastern and Central U.S.
Pallid bat	*Antrozous pallidus*	Western U.S.
Pocketed free-tail bat	*Tadarida femorosacca*	Arizona and California
Red bat	*Lasiurus borealis*	Eastern two-thirds of U.S., plus California and Arizona
Seminole bat	*Lasiurus seminolus*	Gulf Coast
Silver hair bat	*Lasionycteris noctivagans*	All of U.S. except deep southeast
Small foot myotis	*Myotis subulatus*	Western and Northeastern U.S.
Spotted bat	*Euderma maculatum*	Western U.S.
Western big-ear bat	*Corynorhinus rafinesquei*	Western and Central Midwestern U.S.
Western mastiff bat	*Eumops perotis*	California
Western pipistrel	*Pipistrellus hesperus*	Western U.S.
Western yellow bat	*Dasypterus ega*	California
Yellow bat	*Dasypterus floridanus*	Gulf Coast and Florida
Yellow bat	*Dasypterus intermedius*	Gulf Coast and Florida
Yuma myotis	*Myotis yummanensis*	Western U.S.

769. Where do the more common bats nest throughout the year?

The table below gives examples of nine species as reported by the State of Massachusetts, Division of Fisheries and Wildlife office:

Common Name	Species	HABITAT Summer	Winter
Big brown bat	*Eptesicus fuscus*	Buildings, trees	Buildings, caves, mines
Little brown bat	*Myotis lucifugus*	Buildings	Caves, mines
Northern long-eared bat	*Myotis septentrionalis*	Trees, building exteriors, only occasionally inside buildings	Caves, mines

(continued)

Pest Vertebrates Other Than Rodents 113

Common Name	Species	HABITAT Summer	Winter
Indiana bat	*Myotis sodalis*	Caves, mines, hollow trees, beneath tree bark	Caves, mines
Eastern small-footed bat	*Myotis leibii*	Beneath tree bark, only occasionally in buildings	Caves, mines
Eastern pipistrel	*Pipistrellus subflavus*	Trees, rarely in buildings	Caves, mines, rock crevices
Silver-haired bat	*Lasionycteris noctivagans*	Trees	Buildings, trees, rock crevices (migratory)
Red bat	*Lasiurus borealis*	Tree foliage	Migratory
Hoary bat	*Lasiurus cinereus*	Tree foliage	Migratory

770. **Can you summarize bat activity on a yearly basis to help a customer understand why and when they have bats?**

    Using Indiana as an example, bat activity is as follows:
    1. April through May—Re-entering buildings for nurseries (mainly females returning)
    2. June through September—Rearing young and maturing
    3. October through November—Leaving buildings for hibernation sites
    4. December through March—Hibernating

771. **Where are the most common entrance ways through which bats can get into a structure?**

    Bats may enter under vents and screens, under or around porch roofs, through hollow walls, under siding and under roof areas. Roof areas to inspect include drip edge, soffits and fascia boards, ridge caps, under roofing, between house and chimney, and at flashing joints.

772. Which of the common bats in the eastern states commonly invade structures?

    The little brown bat and the big brown bat.

773. What diseases are associated with bats?

    They include rabies, Japanese B encephalitis, histoplasmosis, dermatomycoses, relapsing fevers, leptospirosis and Chagas' disease.

774. Are all the diseases associated with bats carried by those species living in structures?

    No. Histoplasmosis and rabies are the major problems.

775. What special precautions should you take if you go into an attic that has bats?

    Wear a hat, respirator, a long-sleeved shirt and long pants.

776. How long can a bat live?

    Some bats live for more than 20 years.

777. How long can bats go without eating?

    Most species can go only one or two days without food.

778. Why should you wear a respirator when inspecting bat harborage areas?

    It will filter out water droplets in the air that may contain the rabies virus. This is especially important in areas where there are many bats. The respirator will also filter out fungus spores in the bat manure that cause histoplasmosis, a lung disease.

779. What materials are used to bat-proof a structure?

    Use hardware cloth, sheet metal, steel wool, caulking or similar material to pack into cracks and crevices.

780. Why should bats be out of the structure before bat-proofing?

    If bats are stuck inside, they will be unable to escape and will die, creating odor and disposal problems.

781. Why should the bat manure be removed from a structure?

    Besides the odor and caustic properties, manure will attract new bats to an area.

Pest Vertebrates Other Than Rodents 115

782. Are new methods to control bats being investigated?

Yes. Sound devices to repel bats are being investigated. One investigator with the Alabama Cooperative Wildlife Research Unit connected a number of ultrasonic dog whistles to compressed air. After 48 hours of continuous exposure to the noise, a large group of bats left an infested structure. However, non-target animals may have been adversely affected.

783. How are anticoagulants used to control vampire bats?

Diphenadione, an anticoagulant (diphacinone) is injected into cattle. The bat feeding on blood from treated cattle ingests the poison and dies from internal hemorrhage within a few days. In about five days, the anticoagulant is no longer present in the cattle and the cattle are not harmed. A second method is to prepare anticoagulant gel that bats get on their fur from cattle, and ingest when they groom. This method has the advantage that bats can spread the poison to other bats when they return to their roost.

784. Are there any bat repellents?

There are no registered bat repellents.

785. What is currently being done to control bats?

Normally, the pest control operator (PCO) is called to remove just a few bats from a room or attic. Close doors to confine the bat to one room and turn on the lights. Sometimes they can be chased out an open window. A bat that lands can sometimes be caught by quickly covering it with a coffee can or similar container. Release outside after it is determined that nobody has had contact with the bat (otherwise, the bat may be needed for a rabies test). A shot from a CO_2 fire extinguisher will kill the bats, and sometimes they can be knocked down and killed with a broom or tennis racket. Bats are difficult to capture in nets, but sometimes they can be taken with glue boards placed in roost locations or on the end of long poles. There are no toxic materials registered for bat control.

786. Can mouse traps be used to catch bats?

Yes. An expanded trigger mouse trap can be attached to the end of a long stick. The bat is then touched with the trap and may be caught. It is difficult to capture flying bats in this manner, but by waiting a few moments, the bat will usually land and will be much easier to trap.

787. **Is there a best time of year to remove bats and bat-proof a structure?**

    Yes. Using Massachusetts as an example, the bat removal and proofing should take place in May or late summer from August to mid-October, when the fewest number of adult bats should be present. Bats are dormant during the late fall or winter and will not survive release outdoors.

788. **What happens if a structure is bat-proofed in early summer?**

    In early summer, juvenile bats are too young to fly and will hide inside the structure while the adult females feed. If any juveniles are blocked inside, they may invade the living quarters or die, creating odor problems.

789. **What type of foam is good for keeping out bats?**

    Polycol insulating foam will fill the small access holes. It will work its way into these areas and then expand. It can be obtained from local hardware or lumber stores.

790. **If you are successful at building bats out, will they leave the area?**

    Yes, eventually. They may roost in the open under the eaves or on the side of the house, but they will eventually leave the area.

791. **How do you create a one-way exit for bats to keep them from re-entering a structure?**

    If the opening used by the colony is known, fashion a cone of wire and attach the larger end over the outside entrance. Bats will emerge from the narrow end of the funnel, but will be unable to return through the small opening. Another approach involves using a PVC or metal tube attached to the entrance hole, with a plastic or cloth bag over the outside end. The bag is slit to allow bats to leave the building, but no opening is available for them to re-enter.

Pest Vertebrates Other Than Rodents 117

792. Can netting be used to keep bats out?

Yes, some bird netting can be effective in some situations.

793. Do some people consider bats as part of a pest management program?

Yes. Recognize that bats can consume many harmful insects. Many gardening and nature catalogs now feature bat houses that people can put up around their properties to encourage bats to live there. Normally, however, bats will not affect populations of smaller flying insects such as mosquitoes.

794. Where can I get more information on bats and their preservation?

Write to: Bat Conservation International, Inc.
P.O. Box 162603
Austin, Texas 78716
512/327-9721

Moles
(Talpidae)

True or False

795. Moles are rodents.

    False—Moles are insectivores.

796. Moles have very small eyes hidden by fur.

    True

797. Moles have large front feet suited for digging.

    True

798. Moles feed on earthworms, beetles, grubs and other insect larvae.

    True

799. Moles tunnel in lawns looking for food.

    True—They are looking for earthworms, grubs and other food.

800. Pesticides used to control insects in the soil can cause moles to move away from the target area.

    True—But this approach at mole control is less than certain. Sometimes the reduction in food items will cause moles to leave the area. Other times, they can persist on food items unaffected by the pesticides, such as earthworms and some vegetation.

801. Mole burrows are a few inches below the surface of the ground.

    True—They have two types of runways: surface and deep. Deep runways are three to 12 inches below the surface.

802. Moles build volcano-shaped mounds.

True—This is dirt deposited from deep runways.

803. Moles construct special feeding tunnels.

True

804. Moles destroy plants by feeding on them.

False—Plants are destroyed by moles because of their tunneling action. Tunneling exposes roots to water loss. Animals using mole burrows such as deer mice and voles can then do follow-up damage to plants.

805. Repellents are effective in mole control.

False—They only cause the moles to produce more tunnels.

806. Chewing gum is an effective material for mole control.

False—Rumors persist that particular brands of chewing gum (most often Juicy Fruit®) can be removed from the wrapper, rolled up and deposited in mole burrows. It is not clear whether such treatments are supposed to be toxic or repellent to the moles. Scientific efforts by Dr. Robert Corrigan to investigate this did not find any basis for recommending chewing gum against moles.

807. Moles have soft fur. — **True**

808. Two to three moles are found per acre. — **True**—But lawns next to woods or fields may have higher populations.

809. Mole burrows are raised ridges in the surface of the ground. — **True**

810. The common or naked-tail mole (*Scalopus aquatus*) is restricted to the eastern coastal area. — **False**—This mole is found as far west as Texas and Minnesota.

811. The star-nosed mole (*Condybura cristata*) is restricted to the Northeastern part of the United States and parts of Canada. — **True**

812. Star-nosed moles differ from other moles in that they sometimes travel on the surface of snow. — **True**

813. Star-nosed moles make tunnels two to three feet below the ground while other moles do not burrow that deep. — **True**

814. Bulbs chewed near mole tunnels are the work of moles. — **False**—Moles do not eat bulbs. Mice will use the mole tunnels and feed on bulbs.

815. Trapping is the most effective method of mole control. — **True**

816. Trapping is most effective in the spring and fall. — **True**

817. Trapping is most effective by locating the main runways. — **True**

Fill in the Blank

818. Number of mole litters per year (one)
     _____.

819. Number of offspring per litter (three to five)
     _____.

820. Gestation period _____ (about six weeks)
     _____.

821. Age at which moles leave the (about one month)
     nest _____.

Answer the Following

822. How can you best find the main runways of moles for trapping?

    Noted mole authority Dr. Robert Corrigan advises the following:
    1. Look for runways that follow more or less a straight course for some distance.
    2. Look for runways that appear to connect two mounds or two runway systems.
    3. Look for runways that follow fence lines, concrete paths or other manmade borders.
    4. Look for runways that follow the woody perimeter of a field or yard.

823. Is there a way to test to see if a runway is active?

    Sometimes you can poke small holes with your finger in runways at several locations. Moles will repair these in main runways within a day or two. Mark the holes you make with a stick or spray paint. If holes are not repaired over several days, then these runways are no longer active, and trapping at these locations will not be effective.

824. How many traps are needed to trap moles?

    Use from three to five traps per acre.

825. Name two types of traps used to control eastern moles.

The harpoon or prong-and-choker trap.

826. What type of trap is used to control western moles?

The scissors-jaw trap.

827. What are the best types of mole traps and how should they be used?

Harpoon traps are the most practical. Dr. Robert Corrigan best summarized how to trap for moles in the March 1987 issue of *Pest Control Technology* magazine:

Use enough traps! Unless the mole activity is extremely light, more than one trap should be used. Try to place one trap in each of the main runway areas. Moles can be difficult to trap, so set traps with patience and care. To properly set a harpoon trap on a surface run, carefully follow these steps:

1. Using the side of your hand, lightly press down a narrow section (approximately one inch in length) of an active runway so that the runway is collapsed to one-half of its original dimension.
2. Push the supporting spikes of the harpoon trap into the ground, one on either side of the runway, until the trigger pan just barely touches the depressed tunnel. Be sure the trap is centered over the runway and the supporting spikes do not cut into the tunnel below.
3. Set the trap and leave it, taking care not to tread on or disturb any other portion of the runway system.
4. Traps should be checked twice a week. If a trap fails to catch a mole within four or five days, move the trap to another portion of a main runway system.
5. To prevent children, pets or wild animals from tampering or accidentally springing the traps, plastic pails or other objects can be placed over the traps, and marked with the appropriate warning signs.

Pest Vertebrates Other Than Rodents 123

828. Can you trap a mole with a pit trap, instead of buying an expensive trap?

> Yes, although it takes more work to do a careful excavation. A pit trap can be built with a coffee can or bucket. Insert it in an active tunnel by digging a pit through the tunnel, deep enough so that the top of the can meets the tunnel floor. Pack walls of dirt around the rim of the can. Leave a little space between the can and ground surface, and cover the trap with a board or shingle to completely shut out light. A tunneling mole should drop into the can within a day or two if you have chosen an active runway.

Opossums
(Didelphis virginiana)

829. **Where is the opossum found in the United States?**

    Originally it was confined to the Southeast, but now extends west all the way to California and Washington, and north to Michigan.

830. **Is there only one kind of opossum?**

    Yes, only the species *Didelphis virginiana* exists. It is also called the Virginia opossum.

831. **What does the opossum eat?**

    Opossums usually feed on amphibians, birds, eggs, snakes, earthworms, insects, fruit and berries.

832. **What type of habitat does it favor?**

    It favors swamps, along streams and waterways, farmland and wooded areas.

833. **Can it climb?**

    Yes, the opossum is a good climber and often climbs trees.

834. **Where do opossums rest during the day?**

    They find shelter in the unoccupied burrows of other animals, hollow trees or logs, and under rocks.

835. **Are opossums related to kangaroos?**

    Yes, in that both animals are marsupials; they carry their young in a pouch.

836. **Are opossums destructive?**

    They will on occasion kill chickens and smash eggs. They also destroy corn in fields. Opossums may invade garbage cans and carry cat fleas.

837. **What is the opossum life cycle?**

    They have one or two litters per year. At birth, 18 or more are present, but only 13 mammary glands exist, so some young must die. Usually less than 10 survive per litter.

838. How many animals can be in a single den?

   A dozen or more opossums may be present in a den during the late summer.

839. Will an opossum hibernate?

   No, but during very cold weather it may seek shelter for several days at a time to avoid frostbite on its naked tail and ears.

840. What do opossums eat?

   Opossums eat insects, amphibians, birds, eggs, snakes, earthworms, fruit and berries. They also like garbage and are attracted to carrion. Many die attempting to feed on road kill.

841. How long do opossums live?

   Up to eight years. Twelve to 13 days after mating, the young are born. The young move to a pouch in the belly and feed on the mother's milk for four to six weeks.

842. How large is a newborn opossum?

   The young are smaller than a honeybee. They weigh 1/270th of an ounce.

843. How large do the adults get?

   Opossum adults can weigh up to 14 pounds and reach a length of more than three feet, including a 15-inch rat-like tail.

844. How do you control opossums?

   Live trapping is possible. No. 1-1/2 or No. 2 steel traps are also effective in killing opossums.

845. Is there any special bait to use in trapping opossums?

   Bait of meat, fish, moist dog food or chicken guts is effective. Sardines are especially good bait.

Rabbits
(Sylvilagus)

846. What is the Latin name for the eastern cottontail rabbit?

Sylvilagus floridanus. It belongs to the order Lagomorpha, family Leporidae.

847. Are rabbits rodents?

No. Unlike rabbits, rodents belong to the order *Rodentia*. There are superficial similarities but they are not related. There are 18 kinds of rabbits and hares in the United States. Rabbits differ from rodents in that their upper jaw has two pairs of incisors, one located directly behind the front pair. The incisors are completely surrounded by enamel. Rodents have only one pair of upper incisors and one pair of lower incisors, and their enamel is only on the front surface.

848. What is the life cycle of the cottontail rabbit?

There are four to six young per litter and about four litters per year (at least in warmer areas). The gestation period is about 30 days. The newborn weigh three-quarters of an ounce, and are naked and blind. They breast-feed for about two weeks. The female protects the young.

849. How long do rabbits live?

Cottontail rabbits live for five to eight years.

850. What do they eat?

These animals feed on leaves, stems and bark, and on the buds of trees and shrubs, fruits, vegetables, grasses and clover.

851. When do rabbits do most of their eating?

Very early morning and late afternoon.

852. How large are the adults?

They reach a length of 15 inches and weigh two-and-one-half to three pounds.

853. Why are they called cottontail rabbits?

They have a small white bushy tail that can be seen easily as the rabbit runs away.

854. Do they make a nest?

Yes. The female adult will often use body fur to line the nest to keep the young warm. Nests are often found in tall grass.

855. Will they bite people?

Yes. Trying to grab an adult wild rabbit with your bare hands is unwise. Their teeth are quite powerful.

856. What is a jack rabbit?

Black-tailed jack rabbits, *Lepus californicus,* are large rabbits living on open plains in the western states. Adults can reach almost eight pounds. Although they have been introduced into New Jersey and Kentucky, they normally range from Washington south through California, and east to Nebraska, Missouri and Texas. Originally, settlers called them "jackass rabbits" because of their long ears.

857. What is a snowshoe rabbit?

Snowshoe rabbits, also called varying hares, *Lepus americanus,* are small (two- to three-pound) rabbits found in Canada, the Pacific Northwest and New England. Their population fluctuates from year to year in relation to their main predator, the lynx. In other areas, weasels, foxes and mink are among their predators. When populations of snowshoe rabbits become high, it is believed that stress factors may help to limit reproductive success of this species.

858. Do rabbits have parasites?

Yes. They have both internal and external parasites. Rabbits can be infested with fleas, lice, mites and tapeworms.

859. What is rabbit plague?

Tularemia is another name for rabbit plague. It is a disease of rabbits caused by the bacteria *Pasteurella tularensis.* The bacteria is transmitted from rabbit to rabbit by fleas.

860. Are rabbits of any beneficial value?

Rabbits are a main prey species for many predators. They provide sport as small game animals to hunters. The fur of rabbits is used in the manufacture of felt hats. Domesticated rabbits are an important source of meat for many people. In addition, some people keep rabbits as pets.

861. **What signs do rabbits leave?**

    They leave runways, gnawed trees and vegetables, tracks and droppings.

862. **What do rabbit feces look like?**

    They are dark and round. Usually, several of the droppings are found together.

863. **Where is the cottontail rabbit found?**

    It is found throughout most of the United States and can be found in well-populated suburban areas.

864. **Are rabbits protected?**

    Yes, in most states. Therefore, before beginning any rabbit control program, check state and local laws.

865. **What types of damage can rabbits do?**

    Rabbits girdle trees and sometimes even cut down small seedlings. They feed on vegetables and fruits, and over a few nights can destroy a small vegetable patch. Hunters skinning diseased rabbits can get tularemia. Dogs that might catch and eat a rabbit can contract tapeworms.

866. **How can you protect small seedlings from rabbits?**

    Fencing trees will work, but there are more effective ways. Place a cylinder of hardware cloth or mesh screen around the base of each tree. The band should be about three feet high. In areas where the snow is high, the band should be high enough so that the rabbit cannot reach over the top of the band while standing on top of the snow. Repellents are also available.

867. **What types of winter repellents are available to prevent rabbits from attacking trees?**

    There are several commercial repellents available. For best results, follow the directions on the label.

868. How often should the repellent be applied to the trees?

One application in the late fall should be sufficient for the entire winter.

869. How often should you apply the repellent to the garden plants?

Control is effective for about 10 days unless it rains hard. Rains reduce the efficiency of the repellents.

870. How do you treat leafy plants?

Treat leafy plants only upon emergence and until blossoms form. It is not recommended for application to leaves, stems or fruit of plants to be eaten.

871. How is the repellent applied?

By brush or directly sprayed on the plant. The repellent is applied to the bark where the rabbits would chew.

872. Can you live trap rabbits?

Yes, but in the spring and summer they have plenty of food outdoors. Therefore, it may be difficult to do. It may be necessary to pre-bait in a specific area. Set the traps directly in the rabbit runways.

873. What baits should be used to live trap rabbits?

Apples, lettuce, carrots, clover and salt will lure rabbits into a trap. Spring apple cider on the inside of the trap can also help.

874. How far away should you release trapped rabbits?

Rabbits can travel several miles back to their former location. Therefore, trapped rabbits should be taken at least five miles from the trapped area for release. Check local and state regulations to determine if trapped animals can be released. In some states, wildlife relocation is not allowed and the animal must be destroyed.

875. Are there any effective natural rabbit repellents?

Some people claim to have had success with human or dog hair mulch. You may be able to collect such materials from a barber or animal groomer.

Raccoons
(Procyon)

876. **Where are raccoons found in the United States?**

    The common raccoon, *Procyon lotor*, is widespread throughout the country except for portions of Nevada, Arizona and Utah.

877. **What types of habitat do they prefer?**

    Raccoons live near marshes and streams and seek shelter in trees. They rest in hollow trees or logs, rock dens and occasionally burrows.

878. **Do raccoons nest in homes?**

    Yes. They will move right into an unoccupied house, especially the attic area. There are times when they will live in houses where people are still present.

879. **Do raccoons hibernate?**

    In northern areas they do hibernate for the winter.

880. **How long do raccoons live?**

    Up to 13 years.

881. **How big do raccoons get?**

    Adults reach to more than 36 inches and can weigh up to 45 pounds. In general, northern states have larger ones than southern states.

882. **When are raccoons most active?**

    During the night.

883. **How many young are produced per litter?**

    Litters can have from one to eight young, and average four.

884. **When are the young weaned?**

    In 10 to 14 weeks.

885. **How long do the young stay with the female?**

    From one to one-and-one-half years.

886. Can raccoons climb?

They are excellent climbers and are very agile. They can climb up and down pipes in a building and have shown time and again how clever they can be in getting to food stored by humans.

887. What do they eat?

They feed on insects, crayfish, mussels, fish, frogs, birds, eggs, grain, fruits, berries and nuts.

888. Do they wash their food?

Raccoons are sometimes observed appearing to wash their food in ponds or streams. Actually, though, they are searching for food such as crayfish.

889. Are raccoons rodents?

No. They belong to the order *Carnivora*, the flesh-eating animals. They are more closely related to bears, foxes and weasels.

890. Are raccoons a protected animal?

Hunting of raccoons is restricted to certain seasons. In some areas, they are protected.

891. Can raccoons do damage to a house?

If left unchecked, raccoons can destroy a great deal of valuable property. They have been known to destroy rugs, upholstery, dishes and anything else left in a house.

892. Can raccoons remember how to open garbage cans?

Yes, even three months later.

893. How do you control raccoons in a structure?

Live trapping is best. No. 2 double coil spring box traps are used to kill raccoons. New traps have to be boiled in water and natural staining materials before using. Setting a raccoon trap properly is an art that requires experience. The National Fish and Wildlife Service and other sources have bulletins available on methods of setting raccoon traps.

902. **What precautions are necessary before trapping raccoons out of a structure?**

   1. Make sure the young are not inside. This may preclude all baiting during the summer months and late spring.
   2. Find out whether you need a special trapper's license. Some states require going to a special course and taking a written test to obtain this license. Other states only require you to have a trapping permit.
   3. Find out the local and state regulations on releasing wildlife like raccoons. Some areas forbid such releases. The animal has to be killed and buried on the property on which it was caught.
   4. Use gloves and disinfect traps to prevent raccoon roundworm, which may be present in the feces.

903. **What advantages do single-door live traps have over double-door live traps?**

   1. On double-door traps, all the workings for releasing the trap doors are on the exterior of the trap. The animal investigating the trap can touch it and release the doors before it ever had a chance to enter the trap.
   2. By placing the bait in the far back area of a single-door trap, the animal must walk all the way in to reach the bait. This prevents it from using its tail or back end as a wedge to keep the front door from closing (in a properly sized trap for the animal being controlled).
   3. With a self-locking device present on single-door traps, you can leave the trap open while baiting and in releasing the animal.
   4. Newer live traps also have a door stop that will prevent a raccoon from pulling the door inward once inside the trap.

904. **If the trap is on the roof or in an attic, how can you keep the raccoon from damaging shingles or flooring once it is trapped?**

   Use a piece of sheet metal or plywood under the trap. Fasten the trap down.

905. How do you keep the trap from blowing off the roof or tumbling over once the raccoon is caught?

With the use of wire and roofing nails. Attach the trap to the roof. When finished trapping, cover the holes with roofing cement.

906. Why is it better to suspend or lift the bait inside the live trap rather than placing it on the trap floor?

You will get a better "scent line," because the air movement is greater off the ground. It will also lessen the chance of the bait becoming wet and moldy, or being covered with insects or slugs.

907. Any suggestions on how to suspend the bait in the trap?

You can put it inside a small piece of PVC pipe or in an empty film container. Run a wire through the food holder and tie to the top of the trap.

908. Should a trap with a live animal in it be carried by the top handle?

It is best not to, as the animal may reach out while the trap is close to your body. It is better to hold the trap by one end in front of you, or to loop a small piece of wire or chain into the handle and carry it that way.

909. Since raccoons are nocturnal, what is the best time to check the traps?

Check traps early in the morning. This reduces the chance of finding a dead animal in the trap. You should check your live traps every 24 hours.

910. What materials should be taken when baiting for raccoons?

Sardines are excellent bait, but they may also draw cats. If domestic cats are in the area, try jelly, a piece of bread or marshmallows to bait raccoons, because most cats will not be attracted to these.

911. What precautions should be taken when setting a live trap outdoors?

    1. Do not set it on the ground where pets and children can touch it.
    2. Try to hide it as best you can so neighbors and others are not upset at seeing the sight of an animal in a cage.

912. If raccoons are attacking garbage cans and I want to live trap them, any suggestions?

    Knock over an empty garbage can and leave it in place with the live trap inside. The raccoons are so used to crawling into the can, they will often go into the baited trap. Be sure to follow up in the early morning.

913. Are there special chimney caps that will keep out raccoons?

    Yes. Fireplace Technologies has such a device, which allows you to close and open as a damper from within the room. Other firms also sell chimney caps that will keep out raccoons and squirrels, including HYC Co.

914. Once a raccoon is in a chimney, is there a trap to catch it?

    Yes. Target Animal Scent produces a Chimtrap for the humane capture of raccoons.

915. Are raccoons associated with rabies?

    Yes. Along with the skunk and the fox, they rate high as carriers of rabies. From 1992–1993, for example, New York state had a rabies epidemic, particularly in raccoons. Health officials confirmed a record 1,761 rabid animals during 1992 and identified even more in 1993. A total of 1,088 people were treated for rabies exposure in New York during 1992, up from 81 in 1989. In August 1993, an 11-year-old girl died from rabies, the first human death since 1954 in that state.

    The June 1993, rabies summary for Massachusetts showed rabies present in 85 towns. Of the animals tested, raccoons had the highest incidence of rabies. One cat, three skunks and two bats were also confirmed to have had rabies in that year.

916. What other disease is associated with raccoons?

Other than rabies, raccoon roundworm is concern. Of 300 raccoons trapped in the Midwest, a study found 83 percent tested positive for the *Baylisascaris migrans* larvae. These eggs can be viable for 13 to 14 months. They have caused death in a few children. The worms migrate to the brain, where in large numbers they can be deadly. The worms are also associated with woodchucks. If you are handling live traps, be sure to disinfect them.

917. Do raccoons have a specific pattern of defecating?

Yes. They will commonly return to the same area and their droppings will accumulate over many months. Unfortunately, this can be on the roof of a house or in a sandbox.

936. Shrews can eat an amount equal to their weight in 24 hours. True

937. Shrews inject a toxin in mice that paralyzes them. True

Fill in the Blank

938. Number of litters per year _____. (three to four)

939. Number of young per litter _____. (four to six)

940. Length of gestation period _____. (21 to 22 days)

941. Time required to reach adulthood _____. (four to six weeks)

942. Mating age of the shrew _____. (three months old)

Answer the Following

943. What are the major species of shrews in the United States?

    There are four main species:

    Short-tailed shrew —*Blarina brevicauda*
    Long-tailed shrew—*Sorex personatus*
    Water shrew—*Neosorex palustris*
    Least shrew—*Cryptotis parva*

Skunks
(Mustelidae)

944. Are skunks rodents?

    No, they belong to the weasel family, Mustelidae.

945. What do skunks eat?

    They feed on insects, particularly those living in the ground such as grubs and immature yellowjackets. They will eat eggs from ground-nesting birds, as well as frogs, carrion and small rodents.

946. Is there more than one species of skunk?

    Yes, there are several species including:

Common Name	Species Name	Distribution
Common skunk	Mephitis mephitis	Widespread
Hooded skunk	Mephitis macroura	Southern Arizona and Mexico
Hog-nosed skunk	Conepatus mesoleucus	Southwest
Spotted skunk	Spilogale putorius	Throughout much of the United States

947. What is its life cycle?

    In the North, skunks breed in late February and early March. There are four to seven per litter. The female carries them for seven weeks.

948. Where do skunks rest?

    They seek out hollow logs, ground burrows or crawlspaces under structures.

949. How large does a skunk get?

    The common skunk weighs up to 10 pounds.

950. When are skunks most active?

    At night.

951. Do skunks hibernate?

   No. During extreme cold periods they rest, but they do not hibernate.

952. What types of damage can they do?

   Skunks leave holes in lawns while digging for food. They knock over garbage cans in search of food and, when alarmed, release a strong odor. They sometimes invade poultry houses and kill chickens. If not annoying people, they should not be destroyed. They serve a useful purpose in nature.

953. What does a lawn look like that has been damaged by skunks?

   There are small pits three or four inches across. Sometimes large patches of sod are rolled back. The damage occurs during the night.

954. Do skunks carry the rabies virus?

   Yes. They are one of the major vectors carrying the rabies virus. In some states, skunks have become the major animal spreading rabies. Not all skunks carry the virus, however.

955. What happens if you get "hit" by a skunk?

   If it comes in contact with the eyes, the chemical can cause temporary or even permanent blindness.

956. What chemical is released by skunks?

   Methyl mercaptan.

957. From where does the odor come?

   Anal glands house the chemical. The skunk raises its tail and emits fluid up to 10 or 15 feet.

958. How can I reduce the skunk odor in a structure?

   Take wide pans and fill with tomato juice. The tomato juice will absorb the odor. Throw the impregnated tomato juice away. Other deodorants used are neutroleum alpha, a weak solution of vinegar, and household chlorine bleach.

Pest Vertebrates Other Than Rodents 143

959. **How can I get the odor out of a dog?**

    Bathe the dog in tomato juice, or use a mixture of one quart of three percent hydrogen peroxide (from a drugstore), one-quarter cup baking soda and one teaspoon of liquid soap. This solution will quickly remove skunk odor from pets. Rinse the pet well after the bath.

960. **Can Coca-Cola® be used to "de-skunk" a cat?**

    It has been reported that this works using the following method: Soak pet with water, then drench it with the Coca-Cola. Rinse off the cola and check if the smell has come off to your satisfaction. If not, then apply more cola. Once rinsed, apply shampoo for final cleaning of fur and skin. Rinse again and dry.

961. **How can you reduce skunk odor in a window well?**

    Use powdered swimming pool chlorine or regular chlorine bleach. Sprinkle half a cup in the window well. For liquid bleach, use about a gallon.

962. **What is a "civet cat"?**

    Another name for the spotted skunk.

963. **How do you control skunks?**

    There are several approaches used to keep skunks away from particular premises.

    1. Bright lighting at night sometimes works.
    2. Keeping tight lids on garbage cans discourages skunks from invading for food.
    3. The use of soil insecticides depletes the skunk's food supply in the ground.
    4. Plugging holes where skunks enter under a structure is useful in preventing re-entry. Be sure to plug the hole after the skunk leaves for the evening.
    5. Live traps and No. 1-1/2 steel traps are commercially available.

964. Where do you place a trap for skunks?

    A trap is placed near the entrance of its den, but far enough away so that the skunk will not drag itself into the burrow after it is caught.

965. Do you have to bait the trap?

    Sometimes not, but for better success, place the head of a dead chicken or mouse on top of the trap or hang it over the top of the trap.

966. What happens if the skunk is still alive in the steel trap when you arrive on the scene?

    The trap should be attached to a 10-foot pole before it is set. This way, you can dispose of the skunk without getting hit by the odor.

967. How do you dispose of the skunk?

    Drowning or exhaust fumes from an automobile will kill it. Chloroform or carbon disulplhide will also work.

968. Is there some way you can keep the odor confined to the trap area when live trapping?

    Covering the trap with burlap will help. You will have to bait the trap with sardines or cat food.

969. Any safety tips when trapping for skunks?

    1. Put the trap where you can retrieve it easily. You don't want to have to go into a crawlspace to retrieve a trapped skunk. Funnel the skunk to where you want to catch it. Use firewood, cinder blocks or other items to accomplish this.
    2. When removing the trap with a live skunk inside, use an old blanket to cover it. Plastic tarps make too much noise and can trigger the skunks to spray. When covering the trap, use the blanket to serve as a shield between you and the skunk. Skunks can hit with accuracy from eight to 10 feet away.
    3. Where permissible, transport the skunk in an open pick-up truck, not a closed van, just in case it sprays.

Pest Vertebrates Other Than Rodents 145

970. **When live trapping skunks, how many miles away should you take a skunk from the original area before releasing it?**

    If wildlife relocation is allowed in your area (check with local authorities), the skunk should be taken 10 miles or more.

971. **Is there a fence that will keep skunks out?**

    Fences must be three feet above ground and one foot below to keep skunks from climbing over or digging under them. The lower six inches of the wire mesh should be bent outward at a right angle. Use two-inch mesh wire.

Snakes
(Colubridae)

972. **Do snakes lay eggs?**

    Some do. Others give birth to live young. Egg-laying snakes include coral snakes, milk snakes, black rat snakes, king snakes, bull snakes, indigo snakes, red-bellied snakes, ring-necked snakes and hog-nosed snakes. Live bearers include water moccasins, copperheads, timber rattlesnakes, pygmy rattlesnakes, garter snakes and boa constrictors.

973. **Do snakes shed their skins?**

    Yes. The old, shedded skin is often found wrapped around a stick or pipe.

974. **Why are snakes called "cold-blooded"?**

    The body temperature changes with the temperature of the environment. Below 50 degrees Fahrenheit, snakes become inactive, and, in the North, they hibernate during the winter.

975. **What is a cottonmouth?**

    The water moccasin or cottonmouth is a poisonous snake found from Virginia to Florida. It lives primarily in marshes and shallow waterways. The Latin name of this snake is *Agkistrodon piscivorus*. These snakes can live for 20 years or more in captivity.

976. **What are milk snakes?**

    Milk snakes are a kind of king snake with a mottled cream-and-brown skin. They feed primarily on rats and mice, and are often found around barns that are infested with rodents. The snake is beneficial and non-poisonous.

977. **How many different kinds of snakes are there?**

    There are approximately 2,700 different species of snakes worldwide, but only about 116 different species in the United States.

978. **Why does a snake keep sticking out its tongue?**

    The tongue is used to smell the environment and indicates to the snake what environmental conditions exist in front of it.

Pest Vertebrates Other Than Rodents 147

979. Are all snakes poisonous?

    No. Most snakes are beneficial and not harmful to humans.

980. What is the most poisonous snake?

    Many people consider king cobras the deadliest snake because of the potency and volume of its venom. King cobra snakes can be 18 feet long, but are not native to the United States.

981. How do you know which snakes are poisonous in the United States?

    In this country there are 15 species of rattlesnakes, two species of coral snakes and two species of water moccasins, and all are considered poisonous snakes. We can divide the poisonous snakes into two groups: (a) pit vipers (rattlesnakes) and moccasins (copperhead and cottonmouth), and (b) coral snakes. All poisonous pit vipers in the United States have a deep pit on each side of the head between the eye and nostril. The pits are large enough to see at a distance. This leaves the coral snakes, which are colorful and distinct, having alternating bands of red, yellow and black.

982. Where can I obtain a good reference to show which snakes are poisonous?

    The United States Government Printing Office offers an excellent book titled *Poisonous Snakes of the World*. It was published by the Bureau of Medicine and Surgery, Department of the Navy. The manual has excellent full-color photographs of several of the snakes. Contact:

    Superintendent of Documents
    Government Printing Office
    Washington, DC 20402

983. What is the biggest snake in the world?

    The anaconda or water boa of Central and South America. They reach lengths of 30 feet and weigh more than 250 pounds.

984. Must a rattlesnake coil up before it strikes?

    No, although many times they will do so. Sometimes a rattlesnake springs forward without any warning.

148 VERTEBRATE PEST HANDBOOK

985. Can you tell how old a rattlesnake is by the number of rattles it has?

No. More than one rattle is formed annually (every time the snake sheds its skin) and other rattles break off from time to time.

986. Is there any snake that is immune to the poison of the rattlesnake?

Yes. The king snake, also called a chain snake, is immune. Rattlesnakes are part of the diet of many kinds of king snakes.

987. Why do snakes move into a structure?

Snakes enter structures to seek shelter. Moist, cool basements with rock foundations are ideal harborage areas for some snakes. If a rodent population exists in the structure, it makes living conditions all the more favorable. Sometimes the snake enters a structure strictly by accident and may be unable to find its way out.

988. Where are poisonous snakes found?

Rattlesnakes are found throughout much of the United States. Coral snakes are also widespread. Some people are surprised to find that New York state has three species of poisonous snakes, including the timber rattlesnake, the massasauga (a small rattler) and the copperhead.

989. What is the first step in controlling snakes?

Find out what species of snake you are dealing with and a little bit about its biology.

990. How do you control snakes?

Snake control can be approached in several ways:
1. Snake-proofing or mechanical blockage.
2. Removal of food and cover, including trash, debris, rock piles and tall grass that harbor snakes or rodents.
3. Search and kill.
4. Mechanical traps and pit traps.
5. Glue boards (indoors).
6. Repellents.
7. Common sense.

991. How do you remove a live snake from a glue board?

Pour a little vegetable oil on it and hold over a bucket or area of release. You may need to prod the snake with a stick, but the oil will cause the glue to eventually release without harming the snake.

992. How can common sense help control snakes?

By trying to figure out why the snake or snakes first entered the building, you may be able to get them to leave. For example, if a sealed crawlspace (except for one opening) provides a cool, moist area, try blowing hot dry air into the area, and the snake may leave of its own accord.

993. How do you snake-proof a structure?

Openings around foundation walls should be closed. Cellar doors, windows and screens must fit snugly. Snakes can work their way through very small openings, so search carefully. Galvanized screen of one-quarter-inch mesh or smaller can be used to cover drains and ventilators. Yards can be fenced using one-quarter-inch mesh hardware cloth screening, 36 inches wide. The bottom edge should be buried three to six inches in the ground. Vertical fences 24 inches high will keep rattlesnakes out. Supporting stakes should be inside the fence. Keep vegetation from growing near the fence.

994. How do you trap snakes?

Trapping snakes can be accomplished if you can plug up all snake entrances except for one. A 40x32x16 inch wooden trap is placed in front of the exit hole. The trap should have a removable hardware cloth top. The wooden chute with a diameter of four or five inches should open a few inches above the floor of the trap so that snakes do not block the entrance by piling up against it.

A funnel trap with drift fences can also be used. For details, check with your local, state or federal wildlife government agency. Remember that some snakes are on a federal and/or state threatened and endangered species list.

995. What types of pesticides are registered to kill snakes?

   None.

996. What types of snake repellent exist?

   There is no good snake repellent on the market that works consistently. Mercaptan, itself a smelly product, is sometimes used as a rattlesnake repellent, but it may make the snake aggressive. Dr. T's® Snake-A-Way is reported to be effective on copperheads and cottonmouth moccasins. It can be obtained from pest control distributors or from Dr. T's Nature Products Co., Inc.

CHAPTER 8
Non-Toxic Rodent Control Methods

General Considerations

997. How good are cats at rat control?

> A cat is sometimes referred to as an organic rat trap. In general, however, cats make very poor rodent predators. They may take occasional mice, but rats are large and formidable prey for most cats. A survey in India showed that there were more rats in homes with cats than homes without cats; probably because of the additional food source represented by cat food. Also at fault may be the failure of homeowners to apply good sanitation, trapping and rodent-proofing, or having a false belief in the effectiveness of the cat. Cats vary in their interest and ability to catch rats and mice.

998. Are there any dogs that are good rat catchers?

> Most dogs shy away from rats or cannot follow them into the burrow. The rat terrier, however, is considered a good "rat dog."

999. How would one go about setting up a rat survey inspection form?

> The United States Public Health Service has several forms that appear in numerous publications. They may have to be modified or expanded for your specific purposes.

152 VERTEBRATE PEST HANDBOOK

Select A or B. Which is the better choice in each pair for achieving rodent control?

1000.	A—Garbage lid tight on cans	B—No garbage lids present
1001.	A—High grass growing close to building	B—Blacktop surrounds building
1002.	A—Rubbish cleaned up	B—Rubbish piled up in back
1003.	A—Grain spilled over floor	B—Grain in sealed bins
1004.	A—Leaky water pipes	B—No leaks
1005.	A—Stock rotated weekly	B—Stock not rotated
1006.	A—Screen door left open	B—Screen door closed
1007.	A—Windows left open	B—Windows with screens
1008.	A—Open garbage dump nearby	B—Incinerator nearby
1009.	A—Materials stored on the floor	B—Materials stored 12 to 18 inches off the floor
1010.	A—Six-inch-wide painted band around inside perimeter of warehouse, free of obstacles	B—Materials stored flush to walls
1011.	A—Garbage can capacity no more than 12 gallons	B—Fifty-five-gallon drums used for garbage

Answers: You should have picked A for 1000, 1002, 1005, 1010 and 1011. You should have picked B for 1001, 1003, 1004, 1006, 1007, 1008 and 1009.

Indicate with a check which of the following are common signs that rodents have been present:

1012. Black light shows positive glowing spots on floor ___
1013. Grease marks along wall ___
1014. Rodent droppings on floor ___
1015. Teeth marks on edge of door ___
1016. Tail marks in dust on floor ___
1017. Grass beaten down to form distinct trails leading to and away from exterior of building ___
1018. Rodent bait stations dragged from original placement sites ___
1019. Smooth burrows around perimeter of structure ___

Non-Toxic Rodent Control Methods 153

1020. Paper products torn and shredded to form nest-like material ___
1021. Rodent hairs present on window ledge ___
1022. Heavy infestation of rodent fleas found in structure ___
1023. Dead rodents found in basement ___

Answer: All of the above are indications that rodents have been present.

1024. **What can a homeowner do to improve rodent control success in buildings?**
    1. Remove garbage nightly.
    2. Pick up pet food at night; feed pet during the day.
    3. Store all pet food in tight containers, including bird, cat and dog food.
    4. Vacuum old rodent droppings. (Warning: See question No. 339 before vacuuming.)
    5. Do not leave open fruit, candy or food out overnight.
    6. If using wood chips or wood mulch around a structure, keep the layer thin.
    7. Keep shrubs trimmed and not over the roof or touching the building.
    8. Plug holes.
    9. Weatherstrip doors and windows.
    10. Avoid keeping debris like boards and old tires on your property.

1025. **In office buildings, what can pest control professionals suggest to employees that will aid in the rodent control program?**
    1. Do not leave open food in or on a desk. If you keep food at work, store it in tight-fitting plastic, metal or glass containers.
    2. Set up coffee areas in as few spots as possible, and designate them as the only areas to prepare drinks.
    3. Keep items stored off the floor and away from walls.
    4. Report any mouse droppings or other rodent signs to the management or to a designated person.
    5. Before bringing new materials into the building (such as boxes, plants, decorations, etc.), check them at the loading dock or outside the door to be sure they are pest-free.
    6. If outside contractors are doing any work at the premises, insist that they eat in designated areas. Remove garbage daily.
    7. Remove trash from all sources at the end of each day.

8. Make keys readily available to pest control personnel. In particular, they need access to electrical closets, telephone rooms, basement areas, housekeeping closets and pipe chase areas for inspection.
9. Vacuum any rodent droppings so that the pest control professional can know where to concentrate their control efforts when new droppings appear. (Warning: See question No. 339 before vacuuming.)
10. Make it possible for the pest control professional to inspect desk drawers after hours if signs indicate mice are nearby. It takes cooperation at all levels to achieve control.

1026. **What are some tips to keep rodents away from the exterior of buildings?**

1. Trim branches to keep them away from the building. Rodents can climb trees and enter buildings through open windows at the roof level.
2. When selecting shrubbery for around a structure, avoid dense, low-lying plants such as ivy, juniper and pachysandra. These types of plantings offer exceptional harborage for rodents, allowing them to go undetected and making it easier for them to enter the building.
3. Do not pile wood chips too high around a building. Rodents can burrow into them.
4. Avoid any vegetative growth within two to three feet of the exterior perimeter walls.
5. Avoid planting fruit-bearing trees such as crab apples, as these also attract rodents.
6. Keep tight lids on exterior garbage cans and dumpsters.
7. Keep the area around the garbage cans clean.
8. Keep grass cut low.
9. Do not store debris, pallets, boxes and other materials directly on the ground or against buildings.
10. Basement and first-floor windows should be kept closed or properly screened at all times.
11. Seal holes in walls where pipes and wiring enter.
12. Keep sewer trap lids on tightly.

Non-Toxic Rodent Control Methods 155

1027. **How can you prove that mice are active in specific areas?**

The use of powdered talc or chalk on floors will reveal footprints the next day. A black light may detect mouse urine on the floor or other surfaces. A video camera set running in low light during the nighttime hours can often detect mice. Cleaning up old mouse droppings and inspecting for fresh ones is the surest way.

1028. **If mice can survive with no water, then how can liquid bait be effective?**

In hot, dry areas, mice will drink readily. Select such areas to use liquid baits for mice.

1029. **Why is it so difficult to develop a successful rodent repellent?**

Rodents can gnaw objects without using their tongue or swallowing what they chew.

1030. **What is the most common reason rodent elimination is not achieved?**

The true number of rodents present is often underestimated. As a result, the applicator does not apply enough control materials or put those materials in the right places. A few rodents may be eliminated, but the rest keep multiplying and the problem continues.

1031. **How wide an area is needed around the inside perimeter of a facility to provide a pest controller with adequate room to do rodent control?**

A band of 12 to 18 inches is needed to give you room to inspect the perimeter.

1032. **Does bait in a rodent bait station draw the rodents to the area?**

No, they are already present.

1033. **How long can it take trying to get rid of one last rat or mouse in a facility?**

It can take weeks, and, in some cases, can prove nearly impossible, depending upon the situation.

1034. What are the four elements of successful rodent control?

1. Eliminate available food.
2. Remove shelter.
3. Rodent-proof buildings.
4. Use trapping, poisoning or both.

Rodent Stoppage

1035. What is rodent stoppage?

It's the blocking off of all passages by which rodents are likely to enter or leave existing structures.

1036. Why is rodent-proofing so important for successful rodent control?

Poisons give temporary relief, but they do not stop more rodents from entering a building. Altering the environment results in permanent rodent reduction.

True or False

1037. Galvanized sheet metal and hardware cloth are used in preventing rodents from entering a building.

True

1038. Rats and mice can gnaw through wood, aluminum, sheeting and asphalt.

True

1039. When a rat is burrowing and comes in contact with a foundation wall, it will usually burrow down instead of going sideways.

True—This explains why an L-shaped curtain wall is used to prevent rats from burrowing through.

1040. A young rat can squeeze through a hole the size of a quarter (about one-half-inch).

True

1041. A house mouse can enter through a half-inch crack. **True**

1042. Dirt floors in structures are an easy entrance route for rats. **True**

1043. Rats can enter a structure by swimming up pipes into the toilet bowl. **True**

1044. Openings into a structure can be screened with quarter-inch mesh hardware cloth to prevent rodents from entering. **True**

1045. The bottom of wooden doors should have flashing of sheet metal or kick plates and sweeps to prevent rodent gnawing. **True**

1046. Self-closing devices on doors ensure that rodents will not enter via doors. **False**—If people leave a rock or heavy object against the door to prop it open, rodents can still enter.

1047. Openings around conduits and pipes should be fitted with concrete or brick and mortar. **True**—Sheet metal patches will also do the job.

1048. Rodents can enter through floor drains, fan openings and roof vents if these are not properly screened. **True**

1049. Rats can climb the outside of vertical pipes with diameters up to three inches. **True**

158 VERTEBRATE PEST HANDBOOK

1050. Rats can climb the outside of vertical pipes of any size if within three inches of the wall.   **True**—Therefore, rodent stoppage should be done at all levels of a structure.

1051. Foam can be used to keep rodents out.   **True**—But only if it is accompanied by metal mesh.

Answer the Following

1052. What is Stuf-Fit™?

A rodent-proofing material for mice that holds up better than steel wool. It is made of copper so that it will not rust. It can be obtained from your distributor or direct from Allen Special Products, Inc., 800/848-6805.

1053. Is there anything commercially available to prevent rats from entering a structure by swimming up through toilet bowls?

Yes, there is a plastic rodent stopper called J.T. Eaton® Rodent Gard. This item is available from some pest control distributors and from J.T. Eaton Company, Inc.

1054. What determines the size hole a rodent can get through?

The size of the skull. Mice can get through a hole the size of a dime.

Fill in the Blank

1055. _____-gauge galvanized hardware cloth should be used when rat-proofing.   (17)

1056. _____-gauge galvanized hardware cloth should be used when mouse-proofing.   (19)

1057. _____-gauge galvanized sheet metal should be used in rodent-proofing.   (24 to 26)

1058. _____-inch thick brass or   (one-eighth)
       aluminum should be used for
       door kick plates.

1059. _____ mesh hardware cloth   (one-quarter-inch)
       is needed to keep out rodents.

Rodent Odors

1060. What do you do when a rodent dies and starts to smell, and you cannot locate the dead animal?

   There are several commercial deodorizers or masking agents currently used by pest control operators. These include Epoleon® formulations and neutroleum alpha. Some companies are also using ozone-generating machines.

1061. What is Epoleon?

   This product is used to destroy foul odors, including insecticides, dead rodents and skunk spray. The product is available from many pest control distributors and also from Epoleon Corporation of America, 800/376-5366.

1062. Are there any clues to look for in trying to locate the dead animal?

   Yes. Search in the area where the odor is strongest. When present, look for blue bottle and green bottle flies. These are shiny metallic flies that deposit their eggs on dead carcasses. The adult flies will land on a wall or ceiling directly in front of the location of the carcass.

1063. Do rodents leave an odor even when they do not die?

   Yes. The urine is very pungent, particularly with mice.

1064. Should dead rodents be disposed of?

   Yes. Dead carcasses begin to smell within hours, and can spread parasites and disease, or pose a secondary hazard to predatory and scavenging animals (if killed by rodenticides). Do not handle the rodent with bare hands. Wear gloves and seal carcasses in doubled, plastic bags.

1065. Why do some rodents smell worse than others?

> Larger, heavier rodents will smell stronger and longer than smaller ones. The air temperature and amount of air current in the area will also affect the smell, as will the relative humidity.

Traps

Fill in the Blank

1066. Snap traps can be made more effective by _____ the trigger surface.   (expanding or increasing)

1067. Snap traps should be placed _____ to walls.   (perpendicular)

1068. The names of four traps used specifically for trapping mice are _____.   (Ketch-All, Mouse Master®, Tin Cat, Kwik Katch®)

1069. Typical materials used to expand trigger traps are _____.   (cardboard, screen, wire and hardware cloth)

1070. List three different types of traps used to kill rodents _____.   (wooden-based snap traps, steel jaw trap and Conibear trap)

1071. _____ traps are used to capture rodents alive.   (cage or box-type)

1072. The expanded trigger should be about _____ inches square.   (two)

True or False

1073. Before they can be reused, traps must be cleaned in boiling water if a dead rat is found in the trap.   **False**—Clean off rodent hair and feces with a brush or rag.

1074. To make trapping more successful, it is helpful to move boxes and other objects in an attempt to force the movement of the rodent directly into the trap.

True—Use boards, boxes or other objects in the environment to "herd" the rodent to the trap location.

Answer the Following

1075. Does it make any difference which setting is used on expanded mouse traps?

   No, firm and soft settings catch mice with about equal effectiveness.

1076. Are snap traps more effective than glue boards in controlling mice?

   Recent studies at Purdue University show that this is true. However, both still play an important role in rodent control.

1077. Should you use snap traps and glue boards together?

   It is a good idea to do so, because mice that are naturally shy of glue boards will go to a snap trap and vice versa.

1078. Will prebaiting snap traps for rats for a few days greatly increase rat kill?

   Yes, compared to not prebaiting.

1079. Why is this so?

   Without prebaiting, you tend to catch the immature rats and the adults become trap-shy.

1080. Is there a humane mouse trap?

   There have been various traps with this claim. In Sweden, a trap was marketed as humane because it used a pill to knock out the mouse once caught. Woodstream Corp. is currently marketing a new snap trap that is considered humane in some European countries. Snap traps usually kill mice quickly. Live traps and glue boards kill more slowly.

162 VERTEBRATE PEST HANDBOOK

1081. Are expanded trigger snap traps superior to regular snap traps in controlling mice?

    Yes.

1082. Is it critical which trigger setting is used when setting snap traps?

    No, the most critical factor is the placement and position of the trap.

1083. Is there a "best" bait for snap traps?

    No. Different foods and nesting materials work to different degrees, depending upon the environment.

1084. Why is dental floss good for baiting snap traps?

    It is attractive to mice as nesting material. It is also easy to carry in the field.

1085. Are see-through multiple traps less effective than ones with a solid cover?

    No, but they are easier to inspect.

1086. With multiple traps, does it matter which way the traps are facing (parallel or perpendicular to the wall)?

    No. However, setting them perpendicular and flush to the wall when next to exterior doors is more advisable. A mouse entering a new territory may run right past the multiple trap if it has the opportunity to do so. This is possible if the trap is set parallel to the wall.

True or False

1087. Rat traps should be secured so that the rat will not drag itself away and die behind a wall.

    True

1088. Rats can chew off a leg to escape from a trap.

    True—But it will usually die from blood loss.

1089. Using traps is a method of mechanical control.

    True

1090.	Mice are easier to trap than rats.	**True**
1091.	Snap traps are sometimes attached to ceiling beams for the control of roof rats.	**True**
1092.	Traps should be examined daily.	**True**
1093.	In structures where food processing takes place, each bait station should be numbered and dated.	**True**—Preferably, no rodenticides would be used in such areas. Instead, traps can be used inside the bait stations.
1094.	Bacon, peanut butter, fish, ground meat and bread are good baits to use for trapping Norway rats.	**True**
1095.	Nut meats, apples, carrots and bread are good baits to use for trapping roof rats.	**True**
1096.	For mice, one trap every two or three square feet may be necessary.	**True**
1097.	For rats, traps should be placed at least every 20 feet.	**True**
1098.	All rats of the same colony are equal in their tendency to be trapped.	**False**—The young rats and lower-level class adults are easier to trap than the dominant rodents.

Answer the Following

1099. What is a Conibear trap?

> This refers to one particular "quick-kill" type trap. It comes in various sizes and is sold by a number of companies including Woodstream Corp.

1100. Why are there different-sized Conibear traps?

> The higher the size number, the larger the game intended. For example, sizes 110 and 120 are used for muskrats and skunks. Size 220 is used for raccoons.

1101. Are Conibear traps illegal in some states?

> Yes. For example, in Florida this trap was banned along with other leg hold and body grip traps in 1973. Conibear traps can still be used under special permit. In obtaining a permit, the requester has to show that he or she has tried other alternatives without success. Unless the target species is a beaver, the permits are not easy to obtain. Before ordering any type of trap, check with the appropriate local and state agencies to see if it is legal.

1102. Are there any traps other than Havahart's used for live trapping?

> Yes. National Live Trap Corp. manufactures both rigid and folding traps. The folding traps are more convenient to carry in a vehicle. The company has several different sizes in stock and will make special traps on request.
>
> Additional companies selling live animal traps can be obtained from the United States Department of Interior, Fish and Wildlife Service, Atlanta, Ga. 30323. Ask for Wildlife Leaflet 263.

1103. Can you suggest what size of live trap to use in trapping different-sized animals?

> The following suggestions may be followed:
>
> | | |
> |---|---|
> | Mice | 3x3x10 inches |
> | Chipmunks, rats, ground squirrels, gophers, weasels | 5x5x16 inches |
> | Skunks | 7x7x20 inches |
> | Raccoons | 10x12x32 inches |

Non-Toxic Rodent Control Methods 165

1104. How does temperature affect the efficiency of a metal snap trap?

In an extremely cold area like a walk-in cooler, it is important to allow a metal trap to reach room temperature before setting. If you don't, the trap will contract and set itself off after you leave.

1105. When setting snap traps, what must one consider?

There are several factors, including safety to pets and people to consider: (a) if they will be checked daily, (b) if vibrations in the building will trip the traps, or (c) if the traps are set where they are most likely to catch problem animals.

Multiple-catch Traps

Answer the Following

1106. What is a multiple-catch trap?

It is a wind-up trap, such as a Ketch-All, designed to catch house mice.

1107. What is the principle behind the trap?

Mice are inquisitive. They go inside the entrance to investigate. They step on a spring-loaded floor and are thrown into a specific chamber from which they cannot escape.

1108. How many mice can you catch with this type of trap?

Upwards of 20 mice can be caught in one night in one trap. The addition of a special attachment allows the mice to leave the trap and enter a jar filled with soap and water, drowning them. Otherwise, the mice accumulate in the trap and can be dumped out into a bag or bucket for disposal. Some people place glue boards inside the trap to contain the mice.

1109. Will this trap catch rats?

No. Rats are more cautious than mice, and the openings are too small for rats to enter.

1110. Are all mice easy to catch in multiple-catch traps?

No. Younger mice generally are easier to catch than adults.

1111. When placing the trap near a wall, which side should be placed nearest the wall—the side containing the wind-up handle or the opposite side containing the rod?

In either position, someone or something can accidentally push it up against the wall and thereby release the wind action. Sometimes the trap is pushed so hard against the wall that it will not allow the mouse to be flicked even though it entered the trap. Although most applicators place the trap with the entrance parallel to the wall, the manufacturer recommends setting the traps about two inches from the wall, with the entrance nearest the wall.

1112. Are multiple-catch traps dangerous to humans?

Ordinarily, no. However, on older traps, if you wind up the trap and press your finger or hand against the rod protruding from the trap, it might release the tension and bore a nasty path into your hand. The newer models do not present this problem.

1113. Why will mice sometimes be on the premises and not enter the multiple-catch trap?

The mice are not in the vicinity of the trap. They may be living inside a wooden pallet loaded with suitable food and harborage, and will not travel far enough to encounter the trap.

1114. Where can you obtain multiple-catch traps?

Distributors are found throughout the United States.

1115. How many times does it take a mouse to get trap-shy?

With a multiple-catch trap, some mice become accustomed to the trap after one catch and want to keep coming back. You can release such a mouse and keep on trapping it again and again. Other mice, once caught and released, will not go back to the trap.

1116. Can mice escape from a multiple-catch trap?

In an earlier model, young, smaller mice could escape through a gap in the trap panel opening. This has since been corrected.

1117. Can mice get out of Tin Cat traps?

This type of trap has a simple treadle trapdoor like a teeter-totter that drops them down into the body of the trap. Sometimes mice work the see-saw door until they can escape. Newer models, however, allow you to remove the "teeter-totter" and install a glue board.

Glue Boards

True or False

1118. Room temperature can affect the efficiency of a glue board.

True—If it is too warm, the glue will begin to melt and run onto the floor. If it is too cold, the trap becomes less sticky.

1119. There are glue boards made specifically for cold environments.

True—Atlantic Paste and Glue Co., Inc. makes a black glue board that can be used inside freezers. It is called the 24WR5 walk-in refrigerator glue trap.

1120. Rats can, in many cases, pull one foot off a board if it gets stuck.

True

1121. Glue boards are more effective than anticoagulant baits.

False—But they are used to supplement baits, and in areas where no traps or pesticides can be used, they are very valuable.

1122. Putting a cover over a glue board increases the catch for mice.

False—Mice will either jump over low covers, or stop to investigate rather than running and getting all four feet on the board. Covers have the advantage of protecting the glue surface from dust and debris, but research at Purdue University revealed that glue board covers can decrease mouse catches.

Answer the Following

1123. Will mice intentionally cover glue boards with debris in an effort to render them ineffective?

> Some adult mice have been known to do this. They are attempting to keep using their familiar pathways.

1124. How can I overcome the problem of glue board shyness in mice?

> Place an expanded snap trap on the glue board so that the set trap, baited with food or cotton, is stuck in the center. When the mouse attempts to reach the food and is not killed by the trap, it will be caught in the glue. Another strategy is to place the snap trap near the edge of the glue board. If the mouse reaches in and avoids the glue, it still gets caught by the snap trap.

1125. Will a mouse chew its leg off when caught on a glue board?

> Only rarely. If you see a leg left in a glue board, it is more likely that the bone was broken when the mouse pulled away from the leg in its efforts to escape.

Non-Toxic Rodent Control Methods 169

1126. Why will you often find two to four young juvenile mice all caught in one glue board in one night?

   Young mice from the same litter tend to run around and chase each other. They can and will all rush onto the same glue board.

1127. Are there ways to make conventional glue boards more effective?

   Trapping with any device requires a certain amount of skill and can be made more effective through experience. Here are a few techniques to increase glue board catches:

   1. Use paper glue boards and curl into a piece of PVC pipe about three inches in inside diameter and 12 to 18 inches long. This will keep debris and dust from accumulating on the glue board. It also serves as an inviting runway for the rodent.
   2. Add Kitty Malt Chocolate to the glue board. This can be obtained at a local pet shop. The product is produced by Eight-In-One Co. Small pieces of this or of chocolate, nuts or other food that mice are feeding on will also work. Use small amounts of food and place directly in the center of the board. This forces the mouse or rat to stretch out and walk onto the glue board.

1128. Can you outline the pros and cons of using glue boards in controlling rodents?

   The following is adapted from an article I wrote in the July 1992 issue of *Pest Control* magazine.

   1. **Pro:** Glue boards definitely work. When properly used, their success is undeniable.
      Con: Glue boards are not effective when they have dust, flour or dirt on them.
      Solution: Place covers over the glue boards, or put them in PVC pipes or in bait stations. When very windy, use a flexible hose with a bend in it so the air cannot blow straight through.
   2. **Pro:** No rodenticide is needed. In view of public perception, the less pesticide the better.

1132. Where are vibrissae located?

   On the head, front feet and hind feet.

1133. What age are mice when they are most likely to get caught on glue boards?

   The majority of mice caught will be inexperienced juveniles. Sometimes, adult mice are caught when one is chasing another and they will both be found on the board.

Repellents

Answer the Following

1134. Are there repellents sold for rodent control?

   Yes, there are polybutene squirrel repellent materials available from J.T. Eaton. There are spray-on bittering agents that can give some protection against commensal or native rodent chewing for plastic garbage cans and other items. The active ingredient is denatonium saccharide, and the trade name is RoPel (Burlington Industries).

1135. Does RoPel work on other animals?

   Yes, it can help deter dogs or cats from tearing into plastic bags and may keep deer from gardens. It is also highly repellent to people, as you will find if you get any on your hands and accidently get it in your mouth.

1136. What will take away the taste of RoPel?

   An effective way to be rid of the taste of RoPel is to eat chocolate.

1137. Aren't bittering agents also put in rodent baits?

   Yes, but at very low levels. They continue to function as repellents to people (especially children) who may eat the bait accidentally. But the concentration is too low to repel rodents and other animals.

Sound Devices for Rodent Control
Answer the Following

1138. Are sound devices sold for repelling rodents?

Yes, these can be found in catalogs and at some distributors. At present, they are considered relatively ineffective. Ultrasonic sound waves do not penetrate even a sheet of paper, and few rodents travel out in the open. Numerous tests have shown that rodents are not eliminated by these devices, nor by electromagnetic wave generators.

1139. What about playing the calls of rodent predators as a repellent?

There are companies selling tapes of such sounds, calling this method of repelling rodents "biosonics." This involves the use of sounds that mimic living organisms or are actual recordings of such animals. While effective in attracting or repelling some animals, these sounds have not proven useful in repelling rodents.

Black Lights for Rodent Contamination Detection
True or False

1140. Dried rodent urine will fluoresce bluish-white to yellowish-white. **True**—But so will many other substances.

1141. Rodent hairs will fluoresce bluish-white. **True**

Answer the Following

1142. How can you tell if a glowing stain is from urine?

There is a "Brom Thymol Blue Urease Test Kit" available. Place suspected object on the test paper in the kit, moisten with water and cover with a cover glass. If a bluish spot appears after three to five minutes, it is from urine. Such tests do not normally distinguish between rodent urine and the urine of other animals or people.

1143. **Where can you obtain such a kit?**

It's from Spectronics Corp.

CHAPTER 9
Rodenticides

General Considerations

1144. What is the maximum distance to allow between the placing of bait stations around the perimeter of a building or open dump?

> There is no set value. Most product labels indicate a range or minimum distance. It depends upon the terrain, rodent population, areas of activity and a host of other factors. Some health officials ask for at least one bait station every 25 feet. Sometimes this is not enough. Other times it may create unnecessary placements. Certainly, more placements are needed for mice than for rats.

1145. Do you measure the success of a rodent control program by how many rodents you kill?

> No. It is measured by how many rodents are left alive. Rodents are prolific breeders, and can soon replenish their ranks unless a good rodent control program has been developed, including sanitation and rodent stoppage.

1146. In cold winter months, why is it difficult to get rats to enter metal bait stations?

> The metal becomes extremely cold, and the bottoms of a rat's feet are bare. This combination can discourage rats from walking inside metal, or even plastic bait stations in the winter. A strip of cardboard placed inside the station serves as a "rug" and increases the chance of the rodents entering.

1147. How can you prevent rodent bait from freezing to the bottom of metal bait stations?

> Place it on a plastic tray inside the station.

1148. List six major reasons rats may not eat rodenticide bait:
   1. Bait is contaminated with insecticide odors. Keep insecticide products and spray away from rodenticides.
   2. Bait is moldy and old.
   3. Bait is of poor quality.
   4. Poor placement of bait.
   5. Bait contains an acute active ingredient causing quick symptoms. Rodent started to feed, became sick and became bait shy.
   6. Rodents have varied food preferences. Check to see what they are eating on the site. Try a different bait.

1149. Arrange the following materials in order of bait acceptance for rats:

   Fish, corn meal, pellets, paraffin bait
   (See answer, below.)

1150. Arrange the same materials in terms of shelf life:

   Answers:

Material	Acceptance	Shelf Life
fish	first	fourth
corn meal	second	third
pellets	third	second
paraffin bait	fourth (least accepted)	first (lasts longest)

1151. Is it possible to keep grain insects out of grain-based rodent baits?

   Yes. Start with fresh, clean bait in clean bait stations. Change baits regularly (every two to four weeks). Remember, use of insecticides on or around baits may reduce rodent acceptance.

1152. What is "secondary kill"?

   This term means that a chemical kills one animal, and then remains toxic enough to kill a second animal that eats the first animal. Sometimes the chemical is in the blood or liver, and sometimes the bait remains in the stomach of the first animal, so if the stomach is eaten, the second animal can also be poisoned.

1153. For each of the following rodenticide products, indicate whether they are anticoagulant (slow-acting) or acute (fast-acting).

　　A. Contrac _____
　　B. Warfarin _____
　　C. Rozol _____
　　D. Talon G _____
　　E. Zinc phosphide _____
　　F. Generation _____
　　G. Maki _____
　　H. Vengeance _____
　　I. Isotrac _____
　　J. Quintox _____
　　K. Strychnine _____
　　L. Final _____

Anticoagulant: A, B, C, D, F, G, I and L
Acute: E, H, J and K

1154. Is it true that rodenticides should never be used within a food processing area, such as a restaurant or large food plant?

　　Follow the product labels. Most rodenticides, such as anticoagulants, allow use inside such facilities when operations have ceased. Rodenticides could be placed in the evening and picked up the following morning, for example. The concern with rodenticide use in such facilities is to prevent the chance for contamination of food by rodenticide or dead and dying rodents. Generally, rodenticides are limited to the outside or non-food areas of such facilities, and traps or glue boards are used inside.

1155. Whom can I contact for help if a pet may have been poisoned by something I did?

　　If you know the product involved, check the label for emergency assistance numbers. Many companies provide this service free of charge, and you can speak with a technical specialist, veterinarian or toxicologist. Some vets recommend that you carry some standard two percent hydrogen peroxide and try to pour some of this down an animal's throat (for example, cat or dog) that you think may have eaten a pesticide. This should cause the animal to vomit, and can be effective if done within a few minutes of ingestion and before the animal has started to digest the poison.

There is a central source for information that may be useful: the National Animal Poison Control Center, located at the College of Veterinary Medicine at the University of Illinois (Urbana-Champaign, Ill.). This service charges for advice. You can call 900/680-0000 and be charged $20 for the first five minutes, then $2.95 a minute thereafter. The average call takes about 10 minutes. Or you can call 800/548-2423 and have a $30 per case charge placed on your credit card.

1156. **What is the National Animal Poison Control Center?**

The NAPCC is a non-profit service of the University of Illinois. It is the first animal-oriented national poison center in the United States. Since 1978, it has provided advice to animal owners and conferred with veterinarians about poisoning exposures. The NAPCC's phones are answered by licensed veterinarians and board-certified veterinary toxicologists.

1157. **What makes NAPCC different from other poison centers?**

Although human poison control centers receive calls about pets and domestic animals, they are not well-versed in treatment of non-human animals. The NAPCC is staffed by veterinarians and has many years of experience with pesticides, drugs, poisonous plants and other poisonings in pets and domestic animals.

1158. **Why is there a charge for this service?**

The charge supports the work that the NAPCC does, and pays for the staff, computer system and record-keeping. Whether the animal owner dials the 800 or the 900 number, the NAPCC will do as many follow-up calls as necessary in critical cases, and will consult with the animal owner's veterinarian if desired. Sometimes just this reassurance for your customer will save you being charged for many hundreds of dollars in veterinary bills. Oftentimes, the animal was injured or poisoned by something else and you may not be liable.

1159. What information should I have available before calling the NAPCC?

> If you are the applicator, have information concerning the exposure, type of material, time since exposure and the problems the animal is experiencing. Be ready to discuss the animal breed, sex, age, weight and how many animals are involved. If you give this number to the pet or animal owner to call, make sure they have the information handy.
> To discuss this service, you may contact Dr. Louise Cote, NAPCC, University of Illinois, College of Veterinary Medicine, 2001 S. Lincoln Ave., Urbana, Ill. 61801; 217/333-2053.

1160. What is the name of a pesticide commonly used to prevent insect infestation of grain-based rodenticides?

> Malathion at 0.1 percent.

Define each of the following terms

1161. **Rodenticide**—A pesticide used to kill rodents.
1162. **Bait**—A material prepared for the purpose of attraction and consumption (it may or may not be toxic).
1163. **Secondary poisoning**—The potential with some poisons for the first animal eating the toxic bait and dying to be toxic to another animal that eats all or part of the first animal. In eating the carcass, the second animal may ingest enough poison to sicken or kill it.
1164. **Anticoagulants**—Materials that prevent clotting of the blood in warm-blooded animals. Common ingredients for rodenticides.
1165. **Emetic**—Type of material used to induce vomiting.
1166. **Bittering agent**—An additive used to make toxic materials taste bad so that they won't be ingested by people.
1167. **Oral LD50**—The value given for toxic substances so they can be rated against each other in terms of their relative toxicity. The LD50 is the dose level that kills 50 percent of the test group. These laboratory studies generally pump different amounts of toxic materials down an animal's throat. The level needed to kill half of the test group is generally expressed as an amount

180 VERTEBRATE PEST HANDBOOK

in milligrams of active toxicant per kilograms of animal body weight. LD means lethal dose, and so this statistic is used with poisons that are ingested. Those that are inhaled through the lungs would have an LC50 value, standing for the lethal concentration of 50 percent of the test group.

1168. **Tracking powder**—Type of rodenticide formulation made with a dry chalk or dust carrier that is placed in areas where rodents walk. The rodents get the toxicant on their feet and fur, and ingest it when they groom. Tracking powder can also refer to non-poisonous chalk dust or flour used to determine where rodents are active by showing their tracks.

Answer the Following

1169. Bitrex® is a popular taste deterrent put in several commercial rodenticide baits. Will Bitrex stop dogs and birds from eating the baits?

> No. The low levels (10 to 50 parts per million, or ppm) of Bitrex used in commercial rodenticide baits are enough to discourage people from eating the bait. However, the levels are not high enough to deter dogs or birds. Dogs bolt down their food without tasting it very much. Birds do not have much of a sense of taste. In both cases, the high level of Bitrex that would be needed to discourage dogs or birds would also keep rodents from eating the bait.

1170. What is the minimum percentage acceptance of multiple-feeding baits that the Environmental Protection Agency (EPA) will allow opposite an unpoisoned standard test diet?

> The minimum is 33 percent versus EPA meal, which contains unpoisoned corn meal, oats, sugar and corn oil. In other words, one-third of the total amount eaten in a laboratory test by rats or mice must be of the rodenticide bait, and two-thirds (66 percent) can be of the unpoisoned standard test diet.

1171. What is the minimum kill allowed by the EPA in most of the laboratory rodenticide test protocols?

> The minimum kill of a group of rats or mice exposed to the rodenticide being tested is 90 percent. Groups consist of 20

Rodenticides

animals and must be replicated once (total, 40 animals). So in a group of 20, more than two survivors will cause the product to fail the test, unless the product has some other attribute that would justify lower efficacy, such as being exceptionally low in hazard to other animals. The minimum may also be reduced, such as to 85 percent, for baits that are highly weatherable (usually involving the addition of paraffin), which generally reduces acceptability.

1172. **Are there any rodenticides that are currently illegal in the United States?**

Yes. In the past, there were products for rodent control containing DDT, arsenic, ANTU (alpha naphthyl thiourea), thallium sulfate and 1080 (sodium monofluoroacetate). These products were hazardous to other animals and the environment and are no longer available.

1173. **What is 1080?**

This product, also known as sodium monofluoroacetate, is a product of war research. It is extremely deadly, has no antidote, and is definitely capable of causing secondary poisoning. It can kill through dermal absorption as well as by ingestion.

1174. **What is 1081?**

This is a related material, fluoroacetamide. Although less toxic than 1080, it is still extremely hazardous to humans and other animals. It was first registered in 1963 but is no longer available in the United States.

1175. **How do you explain what is going on when bait acceptance appears to be excellent, but poor control is being achieved?**

1. There may be so many rats or mice that they are consuming all the bait, and others never get a chance to eat.
2. You may be using a poison that is not very toxic to the species involved. Diphacinone, for example, is a reasonable product against rats, but is not very toxic to mice.
3. You may be setting the bait placements too far apart. Rodents in between the placements will feed on other food.

4. Your baiting program may not be comprehensive enough. You may be missing entire areas, including suspended ceilings and false floors.
5. New mice are entering in large numbers because the building is not rodent-proof. Construction or cold weather may continue to drive mice inside even though others are being killed.
6. If you are using older anticoagulants, you may have resistance in the population.

1176. **What is going on if I have poor bait acceptance?**

1. There may be too much competing food around. Try to reduce alternative food, by sweeping or cleaning up, or by placing edible materials in rodent-proof containers.
2. If using a fast-acting acute material, you may have created bait-shy animals that will not eat the product again for a while because they got sick from it the first time.
3. You may be using a poor quality bait, or one that has become moldy or contaminated by insecticide fumes. Replace with another type of fresh bait, and see if consumption improves.
4. You may have your placements in the wrong places, where rodents are not finding them or do not feel comfortable enough to stop and feed. Relocate bait placements to active areas.
5. Give your baits time to work. Rats may be shy of new bait and bait containers or stations for several days, or even a week or more. You may want to supplement baiting with traps and glue boards so that you can begin to remove individuals from the population more quickly to increase customer satisfaction. Remember that anticoagulants, once eaten, will not kill the rodents for four to 10 days.

1177. **Why are rodent bait blocks used more often than they once were?**

The trends indicate that blocks are currently approaching pellets in popularity, and will soon surpass them. The reason is that today's bait blocks have more attractive ingredients and less wax. Rats and mice will actually prefer many of the modern bait block products to place-packs in bait stations. Bait blocks are weatherable and it is easy to see if rodents have been feeding on them. One of the biggest reasons that the industry is going to the use of more wax blocks is for reduced liability. Blocks can be wired or fastened inside bait stations and other areas, and cannot be shaken out or carried away by the rodents, as sometimes happens with pellets.

1178. **Are there different ways that bait blocks are formulated?**

Yes. The newer extruded blocks contain wax that is forced under pressure through dyes and then cut off at precise lengths. This results in sharp edges for gnawing and gives both smooth surfaces and rougher surfaces, which may increase attraction. This extrusion process does not generate the neat, cast-in-place methods of the older blocks and may improve acceptance of many inert ingredients. Older methods involved melting wax and imbedding whole or cracked grain into the wax. These had the effect of "cooking" all the ingredients and resulted in blocks that were more variable in size and weight, with all-glossy surfaces. Cast blocks could not easily be made in complex shapes with gnawing edges, while these are easier to achieve with the extrusion process.

1179. **What advantages do bait blocks have over other rodenticide formulations?**

Dr. Robert Corrigan in the September 1990 issue of *Pest Control* magazine gave a good summary of advantages and disadvantages of block baits, as follows:

Advantages

1. Provide stability in damp and wet environments, and various outdoor situations.
2. Offer resistance to insect infestation and degradation.
3. Are not easily scattered out of bait stations the way meal or pellet baits can be.
4. Offer an extra degree of bait security should it be needed (can be fastened down).
5. Provide an excellent formulation for the control of roof rats because blocks can be secured to overhead beams,

pipes, utility poles, trees and other hard-to-reach areas where roof rats are active.
6. Present rodents with an object on which to gnaw, which may in turn attract the rodent to feed.
7. Provide economy in use. Pound for pound, the more popular brands of block baits usually cost somewhat less than bulk pellets and considerably less than place-packs.
8. Present low hazards to some species of non-target wildlife such as birds.

Disadvantages

1. Extra care must be taken when using bait blocks in areas where dogs might discover them. Dogs that are accustomed to eating dog biscuits or chewing on dog bones may eagerly seek out and eat block baits in a similar fashion. Proper application procedures, such as securing tamper-resistant bait stations to the ground, should help prevent such accidents.
2. In some blocks, the grain absorbs moisture and swells if the blocks are exposed to high moisture levels repeatedly or for an extended period of time. Such swelling cracks and breaks apart the block, creating waste and potential hazards in some situations.
3. Unless blocks are secured in some manner, rats will occasionally carry away and hoard chunks of bait.

1180. **How can you prolong the longevity of bait blocks in bait stations?**

There are several steps you can take:

1. Drill holes in the bottom of the bait stations. This allows water from precipitation or hosing down of an area to drain rather than to remain inside the bait station, where it might cause blocks to mold.
2. Fasten block baits off the floor, using stations that come with metal rods for this purpose. Or you can modify older stations by drilling holes and stringing blocks on wire (such as from coat hangers), suspending them just off the floor in the feeding compartment area. Most blocks have center holes to allow for this type of placement.

3. Use a square of Styrofoam in the bait station and place the rodent blocks on top of the Styrofoam.
4. Lift the entire bait station off areas that may flood by using rocks, bricks, boards or other objects.
5. Avoid areas where stations are likely to be flooded, such as low spots or near downspouts.
6. In some situations, you can increase protection for bait stations by covering the tops of them with drum covers, boards, shingles or other objects to help shelter them from precipitation.

1181. How can you reduce the chance of a rodent pulling a bait block out of a bait station?

1. Wire it in place by drilling two holes in the bait station and slipping the bait through an attached wire.
2. Place a drop of glue or adhesive putty on the bottom of the block and press it to the wall or floor of the bait station. CatchMaster's Hercules Putty™ works well for this, as do some thick adhesives and hot glues.
3. Purchase bait stations with rods to hold the bait blocks in place.

1182. Is it true that bait blocks that are slightly moldy may be more attractive to rodents than fresh bait?

Yes, many manufacturers that conducted the required EPA testing to allow a label claim for sewer use found this to be so. The theory is that mold organisms release some simple sugars that make the baits more aromatic and flavorful. However, most applicators would choose to replace moldy blocks to improve their appearance, especially for sensitive accounts or where bait may be inspected by regulatory authorities.

1183. When selecting a rodent bait formulation for a rat burrow, which of these baits is the rat more likely to kick back out: loose pellets, bait blocks or place-packs?

Generally, larger items are more likely to be pushed back out. So place-packs would be most likely, followed by blocks and then loose pellets.

1184. **How can you reduce the chance of the rodent tossing the bait back out?**

   Stuff the bait deep into the burrow using a stick, rod or long-handled spoon. Place a rumpled piece of paper in the burrow. In areas where people or pets may be present, you can further seal the burrow by shoveling dirt into the opening and packing it down. Also, select pellets over other bait formulations.

1185. **What is the difference between bait spillage and bait translocation?**

   Bait spillage is bait that is dislodged from its placement by people or non-target animals. Bait translocation is bait that has been carried away from the place of application by the target animal. Bait spillage is always bad because it exposes bait to non-target animals. Bait translocation can be either bad or good, depending on the circumstances. Sometimes it allows for a better kill if rodents carry the bait back to the nest or to protected feeding areas. But sometimes, like spillage, translocation can lead to contamination or exposure hazards.

True or False

1186. Liquid baits work best when rodents have a shortage of water present in their environment. — **True**

1187. Sugar or glycerin added to liquid baits decreases the chance of them freezing and enhances the attractiveness of the bait. — **True**

1188. Mold inhibitors such as 0.1 percent 2, 3, 5-trichlorophenylacetate are sometimes added to dry baits to prevent mold growth. — **True**

1189. When storing rodent baits, it is important to keep them in a closed, clean container so they do not absorb other pesticide or solvent smells. — **True**

1190.	Once cornmeal baits get moldy or dusty, the chance of rodents eating them is near zero.	True
1191.	Rodent cornmeal baits placed in damp areas are rendered useless after a few days.	True—This is one reason that the most popular ready-to-use rodenticide baits today are pellets and blocks.
1192.	It is illegal to place open trays of rodenticide inside bakeries.	True

Answer the Following

1193. Name four different types of rodenticide formulations that are available.

Pellets, bait blocks, tracking powders and water baits. (There are also meal baits, seed and whole-grain baits.)

1194. There are a lot of rodenticides on the market. I am confused by trade names and formulations. Can you tell me what is currently available and their active ingredients?

Active Ingredient	Trade Name	Formulation	Manufacturer
Brodifacoum	Talon	Pellets, place-packs	Zeneca Professional Products
	Weatherblok	Bait block	Zeneca Professional Products
	Final	Pellets, place-packs, blocks	Bell Labs
Bromadiolone	Contrac	Pellets, place-packs, blocks	Bell Labs
	Maki	Pellets, place-packs, blocks	LiphaTech
Bromethalin	Vengeance	Pellets, place-packs	AgrEvo
	Bromethalin Bait	Blocks	J.T. Eaton
Chlorophacinone	Rozol	Pellets, place-packs, blocks	LiphaTech
Cholecalciferol	Quintox	Pellets, place-packs	Bell Labs
Difethialone	Generation	Pellets, place-packs, blocks	LiphaTech
Diphacinone	Ditrac	Pellets, place-packs, blocks	Bell Labs
	Bait Bitz	Pellets, place-packs, blocks	J.T. Eaton
Zinc Phosphide	ZP	Pellets	Bell Labs

Tracking Powder

Answer the Following

1195. Are tracking powders toxic?

>Generally, "tracking powder" refers to rodenticides formulated in an inert powder that is placed where rodents will walk through it, and ingest the poison when grooming. Also, the term is sometimes applied to inert powders made from chalk or talc that are applied in patches to check for tracks in monitoring for rodents.

1196. How do toxic tracking powders work?

>Their effectiveness is based on the principle that the rodent grooms itself. The presence of the powder on the rodent fur triggers a desire to groom, and they ingest the poison in the process. While rats and mice are considered dirty, they may actually spend up to 20 percent of their time grooming.

1197. Does tracking powder work better on mice than on rats?

>Yes, because mice groom more frequently. However, there are situations where tracking powders are a good choice to control rats: in an area that has abundant food, in instances of bait shyness, or where burrow dusting may be the only practical means of control.

1198. What safety and effectiveness tips should one know before using tracking powder?

>1. Apply thinly.
>2. Avoid windy areas, including where fans function.
>3. Avoid wet areas.
>4. Before treating a void, know where the void leads.
>5. Avoid any surfaces where food is prepared or where rodents can track the powder onto food and food surfaces.
>6. Avoid areas where non-target animals can reach it.
>7. Try to apply as close to the burrow as possible, as grooming is often done in burrows.
>8. Guide the rodent to tracking patches by disguising it as a hidden area. Leaning a box or board works well.

Rodenticides

9. Do not create tracking patches with a hand duster as you would for cockroaches. Too much dust becomes airborne. Sprinkle lightly.
10. Avoid using in suspended ceilings or on high ledges. You could contaminate lower areas.
11. Use separate dusters for rodenticides and insecticides. Insecticidal dusts may repel rodents.

1199. How is tracking power applied?

There are three methods: as an uncovered patch, within a tracking station, within an inaccessible wall or floor void or in outside burrows.

1200. What size should a tracking patch be?

Create long narrow patches against walls. One to two level teaspoons in patches three to five inches wide and two feet long is sufficient, unless label directions differ.

1201. What happens if the dust is too thick?

Paper thin is what you want. Mice can be repelled by thick dusts and may avoid touching it altogether.

1202. Where should you avoid using tracking powder?

1. Commercial food preparation areas (including bakeries and canning facilities). Rodents can contaminate the product.
2. Supermarkets, under gondolas (the bottom shelves where food for display and sale is located). Mice frequent these areas, especially under flour, dog food, candy, birdseed and cereal shelves. The area may seem like an isolated unit, but it is not. Mice move from the floor level up to the food packages and back, and could carry the tracking powder onto the goods above.
3. Suspended ceilings. Normal vibrations can cause dust to fall. When someone lifts a tile to do work, the dust could fall down onto their face.
4. Wet or moist areas. In most situations, moisture will render tracking powder ineffective. If using zinc phosphide tracking powder, the moisture will trigger the release of phosphine gas.

5. Open floor areas. Non-targets, like cats, dogs and even children, can come in contact with the powder.
6. Computer rooms and computer desks. Dust can be sucked into computers and cause damage. Also around "clean rooms" used for assembling electronic equipment or preparing film, and in other sensitive areas.
7. Air ducts. When in use, the dust will move from where you applied it.
8. Under refrigerators. This is a common location for mice to hide. The problem is that the refrigerator fan can blow the powder elsewhere, or someone can move the refrigerator.
9. Pet shops. Wild rodents can track the powder to other animals in the shop.
10. Bakeries. Flour, starch and dough are all light-colored. There is too high a risk of contaminating food without realizing it.
11. Dusty areas. Some accounts produce tremendous amounts of dust on an hourly basis. Factories producing cardboard and lumber companies cutting wood are examples. If absolutely necessary to use tracking powder at these accounts, enclose it in some kind of shelter such as a bait station or section of PVC pipe.
12. Sensitive accounts, including schools, food processing plants, pharmaceutical plants, etc.

Note: The use of tracking powder should not be a first line of defense for rodent control. If you must use a tracking powder, justify it.

True or False

1203. Rodents spend about 20 percent of their waking time grooming. — **True**

1204. Rats groom themselves very thoroughly over a 24-hour period. — **True**

1205. Mice groom quickly. — **True**

Rodenticides

1206. Rodents do not groom each other. **False**

1207. When using tracking powder for rodents, keep it paper thin. **True**—If thicker, some rodents will avoid it.

1208. Even if paper thin, mice tend to be shy of tracking powder. **True**

Answer the Following

1209. What are tracking stations?

   They can be rodent bait stations, or stations made from PVC pipe. PVC pipe sections should be at least 18 inches long and one to two inches in diameter for mice. It should be three to four inches in diameter for rats. Sometimes the only stations that are allowed in an account are tamper-resistant bait stations. The potential hazards of the area will help dictate what you should use.

1210. Are there any hazards associated with using a tracking powder within a wall void?

   Besides the normal wind currents and wondering where the void goes, there is another concern. Electricians and plumbers working in these areas may refuse to work until all the powder has been vacuumed. This could be very costly to the pest control operator (PCO). Still another concern is that rodents could track the powder to a sensitive area, as might happen in a food processing structure.

1211. Most tracking powder labels allow you to treat exterior rodent burrows only when adjacent to a structure. Is there any product that can be used in burrows away from structures, such as the base of a tree in a park or along a bank near garbage dumpsters?

   Yes. Isotrac® is labeled for such use. Unfortunately, it is not scheduled for re-registration.

1212. Are there any tracking powders that hold up better under moist conditions?

   Yes. Ditrac® does but if the area is actually flooded, its efficacy will be impaired.

1213. **Why do tracking powders contain a higher percentage of active ingredient than their respective bait formulations?**

The amount of active ingredient picked up on the surface of the rodent is less than when it is ingested as bait. Therefore, it has to be more concentrated to result in a lethal dose.

1214. **How long does it take a rodent to die after ingesting a tracking powder?**

It depends on the active ingredient. Zinc phosphide takes only a few hours, whereas, after ingestion, the anticoagulants all require about four to 10 days to take effect.

1215. **What tracking powders are currently available for rodent control?**

No new tracking powders have come along since Dr. Robert Corrigan tested five different products in 1989. As published in *Pest Control Technology* in September 1989 the products are:

Trade Name	Active Ingredient	Concentration	Classification and Comments
Anticoagulant			
Ditrac (Bell)	diphacinone	0.2%	Restricted use/rat and mouse –indoor applications –limited outdoor burrow applications
Isotrac (Bell)	isovaleryl	2.18%	General use/rat and mouse –indoor and outdoor
Rozol (LiphaTech)	chlorophacinone	0.2%	Restricted use/rat and mouse –indoor applications –limited outdoor burrow applications
Non-Anticoagulants			
Ridall-Zinc (LiphaTech)	zinc phosphide	10.34%	Restricted use/mouse control –indoor use only
ZP (Bell)	zinc phosphide	10.0%	Restricted use/mouse control –indoor use only –single-dose action

1216. If you use tracking powders, should you stop using other types of rodenticides?

No. The use of bait, together with tracking powder where possible, provides for a more complete kill of a rodent population.

Zinc Phosphide

True or False

1217. Zinc phosphide is a quick-kill material.

True

1218. Prebaiting increases the effectiveness of this material.

True—This is often helpful.

1219. Zinc phosphide is effective in controlling house mice.

True

1220. Rubber gloves should be used when handling the concentrate.

True

1221. Zinc phosphide does not normally create a secondary kill risk.

True—Although there have been reports of cats dying from eating poisoned rats. These may have been due to higher bait concentrations than are now normally encountered.

1222. Zinc phosphide is effective in killing roof rats and Norway rats.

True

1223. Zinc phosphide is registered for indoor and outdoor rodent control.

True—But it is a restricted-use material.

1224. Zinc phosphide is considered more hazardous than anticoagulant baits.

True

1225. Zinc phosphide is a single-dose poison.

True

1226. Zinc phosphide has a strong odor and taste that is unattractive to most animals.

True—However, it may possess the greatest hazard to birds, like chickens, because they do not have much of a sense of taste or smell. Also, birds do not vomit and so the emetic action is lost.

1227. Rats are not repelled by the strong odor of zinc phosphide.

True

1228. Zinc phosphide kills through heart paralysis and gastrointestinal and liver damage.

True

1229. Zinc phosphide kills rodents quickly.

True—Death may occur within a matter of minutes after eating the bait, and rodents may be found out in open areas.

1230. If rats get a sublethal dose of zinc phosphide, they can recover quickly.

True—Such survivors of a sublethal dose may cause them to be bait-shy to zinc phosphide baits in the future.

1231. Zinc phosphide is hazardous to other animals.

True—However, its strong taste and natural emetic action reduces the hazard with most animals.

Fill in the Blank

1232. Zinc phosphide releases a deadly gas in the stomach called _____. (phosphine gas)

1233. Normal concentration in zinc phosphide baits is _____. (one to two percent)

1234. The name of the emetic sometimes added to zinc phosphide baits is _____. (tartar emetic, also known as antimony potassium tartrate)

1235. Oral LD50 value in rats is _____. (average of 40 milligrams per kilogram, with a range of 35 to 48 milligrams per kilogram)

1236. Treatment for zinc phosphide poisoning is _____. (administration of copper sulfate, then an emetic—having the patient drink liquids and avoid fats and oils)

True or False

1237. Zinc phosphide is a restricted-use pesticide. — True

1238. Zinc phosphide is registered for rat and mouse control. — True

1239. Shelf life for zinc phosphide is about three years. — True

Answer the Following

1240. What are the early zinc phosphide poisoning symptoms in humans?

> Nausea, vomiting, abdominal pains, chest tightness, diarrhea, chills, smell of phosphine on the breath.

to dogs than was originally believed, which has reduced the popularity of retail forms of the product, and makes correct placement of baits especially important. Although not as fast-acting as some other acute products, published reports of bait shyness with cholecalciferol are known. It should not be used repeatedly against the same pest population.

Anticoagulants—General Considerations
True or False

1253. Anticoagulants are commonly used to control Norway rats, roof rats and house mice.

True

1254. Liquid anticoagulants are formulated in salt form and are water soluble.

True

1255. Anticoagulants are not available as block bait.

False—Anticoagulant rodenticides are commonly available as meals, grain baits, pellets, place-packs, blocks, tracking powders and liquid baits.

1256. Vitamin K is the antidote for all anticoagulants.

True—The most common form is the injectable K_1 variety, which has a rapid action. For follow-up and long-term therapy, tablets containing K_3 are sometimes prescribed by the veterinarian or physician.

1257. Anticoagulants inhibit the production of clotting factors (prothrombin) in the liver.

True

Answer the Following

1258. List eight advantages of using anticoagulants (versus acute baits):
    1. Limited hazard to non-targets.
    2. True antidote available in cases of accidental poisoning of people or pets.
    3. Good bait acceptance (present at very low levels in bait).
    4. Rodents die slowly, four to 10 days after eating, so no bait shyness.
    5. Easy to use; especially for the newer single-feed varieties.
    6. Good shelf life.
    7. Economical.
    8. Effective on both rats and mice.

1259. What is the history of anticoagulant rodenticides?

    Dicumarol was first synthesized by a German chemist around 1900. At that time, no one knew of the potential uses for the chemical. In 1922, cattle in Canada feeding on spoiled sweet clover developed hemorrhages. Teams of scientists working on the problem identified the cause as dicumarol. This product was too slow-acting to kill rodents, so a search for a better product was started. More than 300 compounds similar to dicumarol were tried before warfarin was identified.

1260. How did warfarin get its name?

    The Wisconsin Alumni Research Foundation (WARF Institute of Madison, Wis.) found one compound to be the most effective in killing rats. The product was named "warfarin" and was patented in 1941.

1261. When did warfarin come into the U.S. market as a rodenticide?

    It was in 1949, nine years after its discovery, that warfarin bait came into public use for rodent control in the United States. Researchers found that in order for warfarin to be effective, rodents needed to feed on the compound over several days. The modern era of anticoagulant rodenticides was born.

rats, only one gene may be involved. Resistant rodents in Europe may have different mechanisms of resistance than those in the United States. More research is needed to fully understand the phenomenon.

1268. **When was anticoagulant resistance first reported?**

In 1958, resistance was identified and reported about rats in Scotland. Since then, it has been reported in many parts of Europe, including Denmark, England, Wales and the Netherlands.

1269. **When was resistance first reported in the United States?**

During the summer of 1971, resistant rats were suspected on a farm near Smithfield, N.C., by local applicators, including S.G. Flowers, who reported the situation to Dr. Jim Wright of the University of North Carolina. Wright contacted noted rodent authority Dr. William Jackson of Bowling Green State University, Bowling Green, Ohio. Jackson and his student at the time, Dale Kaukeinen (presently with Zeneca), collected some rats and took them to Ohio for laboratory testing. The results established warfarin resistance in the United States, as published in *Science* magazine in 1972.

1270. **Should I be concerned about anticoagulant resistance today?**

Not really; it is just proof that pest rodents can survive our control attempts in surprising ways. But when it was first discovered, warfarin resistance caused great concern and several laboratories were set up to identify what they called "super rats." Applicators and manufacturers scrambled to find rodenticides that could still be used effectively on these resistant rodents.

1271. **What has happened to reduce concerns about resistance to the first-generation (multiple-feed) anticoagulants?**

As a result of documenting anticoagulant resistance, several new products were developed, and, today, anticoagulant resistance is less of a concern. Some single-feeding anticoagulants have the ability to kill warfarin-resistant rats and mice, as do any non-rodenticide products such as zinc phosphide, cholecalciferol and bromethalin baits.

Rodenticides 203

1272. **If rats and mice are resistant to warfarin, will they be resistant to the newer anticoagulants as well?**

    The latest *Mallis Handbook* (1997) recommends that where resistance is suspected, applicators should use only those anticoagulants for which no resistance has yet been demonstrated. In the United States, these would be products containing the actives brodifacoum and difethialone (Talon®, Final®, and Generation™). Rodents have proven to be resistant to bromadiolone and other anticoagulants in some locations in Europe and the United States.

1273. **Do other animals show resistance to anticoagulants?**

    Yes. In fact, the phenomenon was first discovered in humans. Anticoagulants are often used in medical treatments to prevent blood clots and for other conditions, and researchers have found some families with members who are resistant to normally effective levels of anticoagulant medicines.

1274. **How are rodents tested for anticoagulant resistance?**

    There are different published methods that laboratories can follow. One approach involves a no-choice feeding test for different periods according to the species. The anticoagulants in such testing are normally made up at one-fifth strength so that the test is more sensitive. Another approach being promoted in England involves injecting rats with known amounts of anticoagulants and then measuring the clotting response time of their blood with instruments.

    Crude tests can be conducted by applicators by simply holding wild rats or mice in a cage and feeding them the amount of bait that the bait manufacturer or published literature indicates is normally lethal. If they eat the bait and are alive after 10 to 15 days, they may be resistant.

1275. **Aside from possible resistance, what are other drawbacks to using anticoagulants?**

    Although they are the most popular rodenticides used today, anticoagulants are slow-acting (killing four to 10 days after the rodent ingests a lethal dose). This may be a disadvantage where quick clean-out is important, such as sites with a short window for rodent killing (ships in port) or where rodents may be carrying disease organisms. All anticoagulants have the tendency to accumulate on the rodent's body and that of any predator or scavenging animal that eats the rodent. For this reason, anticoagulant baits are not generally used against non-commensal rodents or away from buildings so that the potential for secondary hazard is reduced.

1276. What are the benefits of anticoagulants?

1. They are available in the broadest range of formulation types, including some highly acceptable and weatherable products.
2. They have generally broad labels and are not as restricted.
3. They kill slowly, so bait shyness does not develop.
4. They have a readily available antidote (vitamin K).
5. When used according to the label, they pose little hazard to non-target organisms.
6. They are economical to use, particularly the single-feeding variety.

Baiting at Dumps

True or False

1277. The most important time to bait for rats at a dump is in late fall.

True—This will coincide with the most activity and may prevent a breeding peak; monthly baiting may provide the most consistent rat suppression.

1278. Rats in dumps are found near buried material.

False—Rats live near the active edge of the dump, where they forage on new material brought in.

1279. To estimate the quantity of bait to use, one rule of thumb is to figure on one pound of bait for every linear yard of dump face.

True—However, the rodent population should be considered. It is better to overbait than to not have enough bait.

1280. The sloping face of an embankment should be baited by placing the bait in and around trash items.

False—Bait burrows and, where possible, use tamper-resistant bait stations.

1281. If possible, it is better to close the dump to the public before baiting.

True—To reduce hazard and limit disturbance, close the dump to additional dumping for a few days. Be sure to have bait present, or rats may migrate elsewhere.

1282. Bulldozing operations can continue while using a quick-acting rodenticide at a dump.

False—The dump area should be quiet and undisturbed to give rats a chance to feed, regardless of the type of rodenticide used.

1283. The best time of day to bait a dump is early morning.

False—Bait in the late afternoon so the fresh placements are available when rats come out to forage at dusk. This also reduces exposure to day-feeding scavengers like gulls.

1284. Do not bait a dump when it rains or snows, or when precipitation is expected within 48 hours.

True

CHAPTER 10
References for Training

Videos

1285. Can you recommend any good videos for training personnel on the biology and control of rats?

> Yes, a classic video on rat biology is *Ratopolis,* produced in 1973 by the National Film Board of Canada. Available from Video Development Services, Inc., P.O. Box 701067, Houston, Texas 77270.
> Other videos available from the same source include:
>
> *Biology of the House Mouse*
> *Control of the House Mouse*
>
> You may also contact the National Pest Control Association (NPCA) at 800/678-6722 or the American Institute of Baking (AIB) at 800/633-5137.

1286. Are there any good videos available on other aspects of vertebrate pest control?

> Yes, a two-hour video on skunk control is available from:
>
> Ron Erickson
> Target Animal Baits
> P.O. Box 5345
> Glendale Heights, Ill. 60139
>
> Ron Erickson has other videos on trapping beaver and on trapping basics.

Cassettes

1287. Are there audio tapes available for learning about vertebrate pest control?

>Yes, you can order the following:
>
>>Rodents—Norway Rat—Tape 4
>>Vertebrates—Mice, Skunks and Bats—Tape 5
>
>Each tape is about 36 minutes long. They are condensed reviews of biology and control. Order from the author:
>
>>Dr. Austin Frishman
>>30 Miller Road
>>Farmingdale, N.Y. 11735

Slides

1288. Where can I purchase 35 mm slides on vertebrate pests for use in training?

>They are available from:
>
>>National Pest Control Association
>>8100 Oak St.
>>Dunn Loring, Va. 22027
>>800/678-6722
>>fax 703/573-4116
>
>You can also write for a slide catalog from
>
>>Van Waters and Rogers
>>P.O. Box 34325
>>Seattle, Wash. 98124-1325

Published Material

1289. What books are available on vertebrate pest control?

>A good single source that includes non-rodent wildlife species is:
>
>>*Prevention and Control of Wildlife Damage* (1994)
>>91 chapters, in book or CD-ROM format

Order from:

University of Nebraska
202 Natural Resources Hall
Lincoln, Neb. 68583-0819
402/472-2188

This source covers everything from elk to turtles and includes a special section on registered vertebrate pesticides.

Dr. Robert Corrigan has an extensive chapter on pest rodents in the newly revised *Mallis Handbook*, which also features a chapter on native rodent problems and solutions.

Corrigan, R. M. 1997. Rats and Mice. Chapter 1 in: Mallis, A., *Handbook of Pest Control*, 8th Edition, GIE Publishers.

These two books, published in England, are excellent references:

Buckle, A. P. and R. H. Smith, editors. *Rodent Pests and Their Control*, CAB International, Oxon UK, 405 pages, 1994.

Meehan, A. P. 1984. *Rats and Mice—Their Biology and Control*, Rentokil Library, E. Grinstead UK, 383 pages.

Books and other printed materials are also available from:

National Pest Control Association
8100 Oak Street
Dunn Loring, VA 22027
800/678-6722
fax 703/573-4116

1290. **Is there a good wildlife control journal I can subscribe to?**

Wildlife Control Technology is published six times a year. It can be ordered from:

Wildlife Control Technology
P.O. Box 5204
Department E
Glendale Heights, Ill. 60139

CHAPTER 10
References for Training

Videos

1285. Can you recommend any good videos for training personnel on the biology and control of rats?

>Yes, a classic video on rat biology is *Ratopolis*, produced in 1973 by the National Film Board of Canada. Available from Video Development Services, Inc., P.O. Box 701067, Houston, Texas 77270.
>
>Other videos available from the same source include:
>
>*Biology of the House Mouse*
>*Control of the House Mouse*
>
>You may also contact the National Pest Control Association (NPCA) at 800/678-6722 or the American Institute of Baking (AIB) at 800/633-5137.

1286. Are there any good videos available on other aspects of vertebrate pest control?

>Yes, a two-hour video on skunk control is available from:
>
>Ron Erickson
>Target Animal Baits
>P.O. Box 5345
>Glendale Heights, Ill. 60139
>
>Ron Erickson has other videos on trapping beaver and on trapping basics.

Cassettes

1287. Are there audio tapes available for learning about vertebrate pest control?

> Yes, you can order the following:
>
> > Rodents—Norway Rat—Tape 4
> > Vertebrates—Mice, Skunks and Bats—Tape 5
>
> Each tape is about 36 minutes long. They are condensed reviews of biology and control. Order from the author:
>
> > Dr. Austin Frishman
> > 30 Miller Road
> > Farmingdale, N.Y. 11735

Slides

1288. Where can I purchase 35 mm slides on vertebrate pests for use in training?

> They are available from:
>
> > National Pest Control Association
> > 8100 Oak St.
> > Dunn Loring, Va. 22027
> > 800/678-6722
> > fax 703/573-4116
>
> You can also write for a slide catalog from
>
> > Van Waters and Rogers
> > P.O. Box 34325
> > Seattle, Wash. 98124-1325

Published Material

1289. What books are available on vertebrate pest control?

> A good single source that includes non-rodent wildlife species is:
>
> > *Prevention and Control of Wildlife Damage* (1994)
> > 91 chapters, in book or CD-ROM format

Order from:

University of Nebraska
202 Natural Resources Hall
Lincoln, Neb. 68583-0819
402/472-2188

This source covers everything from elk to turtles and includes a special section on registered vertebrate pesticides.

Dr. Robert Corrigan has an extensive chapter on pest rodents in the newly revised *Mallis Handbook,* which also features a chapter on native rodent problems and solutions.

Corrigan, R. M. 1997. Rats and Mice. Chapter 1 in: Mallis, A., *Handbook of Pest Control,* 8th Edition, GIE Publishers.

These two books, published in England, are excellent references:

Buckle, A. P. and R. H. Smith, editors. *Rodent Pests and Their Control,* CAB International, Oxon UK, 405 pages, 1994.

Meehan, A. P. 1984. *Rats and Mice—Their Biology and Control,* Rentokil Library, E. Grinstead UK, 383 pages.

Books and other printed materials are also available from:

National Pest Control Association
8100 Oak Street
Dunn Loring, VA 22027
800/678-6722
fax 703/573-4116

1290. **Is there a good wildlife control journal I can subscribe to?**

Wildlife Control Technology is published six times a year. It can be ordered from:

Wildlife Control Technology
P.O. Box 5204
Department E
Glendale Heights, Ill. 60139

Another publication is called *Animal Damage Control*, and is published six times a year. Order from

Animal Damage Control
P.O. Box 224
Greenville, Pa. 16125

1291. **What manufacturers of rodent control materials have educational materials available?**

You can obtain some excellent training and reference materials (in addition to product labels, material safety data sheets and catalogs) from these manufacturers (usually free of charge):

Bell Laboratories, Inc.
3699 Kinsman Blvd.
Madison, Wis. 53704
608/241-0202
fax 608/241-9631

LiphaTech, Inc.
3600 West Elm St.
Milwaukee, Wis. 53209
414/351-1476
fax 414/351-1847

Zeneca Professional Products
1800 Concord Pike
Wilmington, Del. 19850
302/886-1000
fax 302/886-1644

1292. **Where can I obtain catalogs on trapping of wildlife?**

The following trap and lure companies are some sources:

Critter Control 800/451-6544
Woodstream Corp., Lititz, Pa. 800/800-1819
Kness Manufacturing Co., Albia, Iowa 800/247-5062
Tomahawk Live Trap Co., Tomahawk, Wis. 715/453-3550
On Target Animal Baits, Glendale Heights, Ill. 708/858-4895

Training Courses

1293. Are there any short courses on nuisance wildlife control?

>   The University of Kentucky offers these conferences periodically. Contact:

>>   Dr. Tom Barnes
>>   Department of Forestry
>>   University of Kentucky
>>   Lexington, Ky. 40546
>>   606/257-7597

>   Purdue University has a Wildlife Nuisance Correspondence Course. Contact:

>>   Continuing Education Business Office
>>   Purdue University
>>   1586 Stewart Center, Room 110
>>   West Lafayette, Ind. 47907-1586
>>   765/494-7209

Index

Numbers in Index refer to question numbers.

Agkistrodon piscivorus, 975
albinos, 103, 261
Alexandrine rats. *See* roof rats
altricial, defined, 5
Amblyomma americanum, 399
animal control sticks, 900
anticoagulants, 668, 783, 1164, 1253–1276
antimony potassium tartrate, 1234
ANTU (alpha naphthyl thiourea), 1172
Apars, 709–729
armadillos, 709–729
arsenic, 1172
arthropods. *See* parasites

bait, 708, 910, 1162
　acceptance of, 1148–1149, 1175–1176, 1183–1184
　bait blocks, 1177–1182
　bait stations, 1032, 1144, 1146–1147, 1180–1181
　cornmeal baits, 1190–1191
　at dumps, 1277–1284
　grain-based, 1151, 1160
　liquid baits, 1028, 1186–1187
　for mice, 283, 294
　monitoring, 198
　multiple-feeding, 1170
　for rats, 121, 157, 179–180, 1094–1095, 1183–1184
　shelf life, 1150
　spillage *versus* translocation, 1185
　storing, 1189
barn rats. *See* Norway rats
bat-proofing, 779–781, 787–792
bats, 730–794
Baylisascaris migrans, 916
birds, and Lyme disease, 406, 462
Bitrex, 1169
bittering agents, 1137, 1166
Black Death, 219, 321, 488–500, 859
black lights, 1140–1143
black rats, 200–240, 1222
black squirrels, 630

black-legged ticks, 398
Borrelia burgdorferi spirochete, 396
Brom Thymol Blue Urease Test Kit, 1142–1143
bromethalin, 1242–1247
brown rats. *See* Norway rats
brush rats, 65, 341
bubonic plague, 498

Cadassous, 709–729
cannibalism, 140, 245
capsicum, 646
capybara, 66
carbaryl granules, 465
castor-oil plants, 601
cats, 429, 960, 997
cave rats, 65, 341
CDC (Centers for Disease Control and Prevention), 347–348, 354, 427
Census bait, 198, 295
chimney caps, 913
Chimtrap, 914
chipmunks, 636–637, 641, 645, 658, 660–683
Chiroptera, 730–794
chlorpyrifos, 465
cholecalciferol, 1248–1252
civet cats, 962
Colubridae, 972–996
commensal rodents, 2, 51
common rats. *See* Norway rats
Conibear traps, 1099–1101
cornmeal baits, 1190–1191
cotton rats, 341, 506–518
cottonmouths, 975
Coypu, 578–587
Cynoff, 465
cypermethrin, 465

Damminix, 468–471
Dasypodae, 709–729
DDT, 1172
deer mice, 9, 341, 352, 358, 519–541
deer ticks, 398, 404, 411–413, 425, 460–461, 467, 474
deltamethrin, 465
Demand, 465
Demon, 465
diastema, 11, 60, 71–72
Didelphis virginiana, 829–845
diseases. *See also* specific diseases

　and bats, 773–774
　deaths from, 329
　and deer mice, 521
　and dog ticks, 474
　and lice, fleas, and mites, 330
　and parasites, 314–330, 395–505
　and raccoons, 915–916
　and shrews, 933
　transmitted by rodents, 331–394
　and white-footed mice, 521
dog ticks, 474
dogs, 435, 441, 959, 998
dormancy, 50
Dursban, 465

ectoparasites, 327
emetics, 1165
endemic typhus, 317, 481–487
EPA, 1170–1171
epidemic typhus, 487
Epoleon, 1060–1061
erythema chronicum migrans (ECM), 408
Euphorbia lathyrus, 601

favus, 306
feral animals, 3
field mice, 6–7, 542–562
fleas, 330, 497
Florida water rat, 63
fluoroacetamide (1081), 1174
flying squirrels, 631
foam, for bats, 789
fox squirrels, 633–634

gas cartridges, 704–705
Giant Destroyer, 704
glue boards, 659, 991, 1076–1077, 1118–1133
Gopher Patrol, 601
gophers, 10, 588–603
gopher's spurge, 601
grain-based baits, 1151, 1160
ground squirrels, 632
groundhogs, 684–708
guinea hens, 437

hamsters, 50
Hantavirus Pulmonary Syndrome (HPS), 333–335
hantaviruses, 320, 331–358, 517, 540

212 INDEX

Haverhill fever, 307, 315, 373-381
histoplasmosis, 307
house mice, 79, 241-313, 342, 371, 434
  activity periods, 280
  anticoagulants for, 1253
  baiting, 283, 294
  inspection areas, 312
  multiple-catch traps for, 1106-1117
  resistance to anticoagulants, 1267
  signs of, 295
  supermarket locations, 313
  trapping, 284-285, 289, 294, 310, 1106-1117
  and zinc phosphide, 1219
human granulocytic ehrlichiosis (HGE), 447, 473-480
humane mouse traps, 1080
Hymenolepis nana, 389

infectious jaundice, 318, 359-366
Ixodes dammini. See deer ticks
Ixodes pacificus, 398
Ixodes scapularis, 398

jack rabbits, 856
J.T. Eaton's 4 The Squirrel Repellent, 644

Ketch-All traps. See multiple catch traps

land tortoises, 595
Leptospira ballum, 359
Leptospira icterohaemorrhagiae, 359
leptospirosis, 318, 359-366
Lepus americanus, 857
Lepus californicus, 856
lice, 330
liquid baits, 1028, 1186-1187
live trapping, 1071, 1102-1103
  chipmunks, 667, 678, 680
  rabbits, 872-874
  raccoons, 893-897, 903, 908-909, 911-912, 914
  skunks, 968, 970
  squirrels, 648-649, 651-652, 655
  woodchucks, 707
lone star ticks, 399
lowland marmots, 684-708
lures, 901
Lyme disease, 322, 395-472, 480. See also ticks
  Borrelia burgdorferi spirochete, 396

and Damminix, 468-471
hosts for, 422-423
and house mice, 434
and *Peromyscus* species, 541
prevention, 414, 451, 472
rodents associated with, 422-423
vectors for, 433, 459, 463-466
and white-footed mice, 422, 426
lymphocytic choriomeningitis (LCM), 297, 325, 367-372

malaria, 324
Malathion, 1160
Marmota flaviventris, 686
Marmota monax, 684-708
meadow mice/voles, 6-7, 542-562
Mercaptan, 996
metabolic water, 278
meta-populations, 272
methyl mercaptan, 956
mice, 268. See also house mice
  deer mice, 9, 341, 352, 358, 519-541
  droppings, 58, 78
  meadow mice, 6-7, 542-562
  multiple catch traps, 1085-1086, 1106-1117
  nesting, 75
  orchard mice, 8
  resistance to anticoagulants, 1267, 1272
  trapping, 1068, 1090, 1096
  white-footed, 422, 426, 519-541
Microsorex hoyi, 924
milk snakes, 976
minimum kill, 1171
mites, 330
mold inhibitors, 1188
moles, 53, 795-828
mothballs, 642
mountain rats, 65, 341
multiple-catch traps, 285, 1085-1086, 1106-1117
murine plague, 219, 321, 488-500, 859
murine typhus fever, 317, 481-487
Mus domesticus. See house mice
Mus musculus. See house mice
muskrats, 563-577
Mustelidae, 944-971

naked-tail moles, 810
National Animal Poison Control Center, 1155-1159

Nature's Farewell, 601
nematodes, 159
Neofiber alleni, 63
Neotoma genus, 65
netting, 792
non-commensal rodents, 506-708
  chipmunks, 660-683
  cotton rats, 506-518
  deer mice, 519-541
  muskrats, 563-577
  nutria, 578-587
  pocket gophers, 588-603
  porcupines, 604-617
  prairie dogs, 618-624
  squirrels, 625-659
  voles, 542-562
  white-footed mice, 519-541
  woodchucks, 684-708
non-rodent pests, 709-996
  armadillos, 709-729
  bats, 730-794
  moles, 795-828
  opossums, 829-845
  rabbits, 846-875
  raccoons, 876-917
  shrews, 918-943
  skunks, 944-971
  snakes, 972-996
non-toxic control methods. See rodent control, non-toxic
Norway rats, 80-199, 201, 207, 212, 215, 222, 325, 342, 670
  activity periods, 104, 122
  anticoagulants for, 1253
  baiting, 121, 157, 179-180, 1094
  controlling, 176
  disposing of, 128
  habitat, 82, 128, 226
  locations, 87, 199
  trapping, 107, 157, 1094
  and zinc phosphide, 109, 1222
nutria, 578-587

Ondatra zibethica, 563-577
opossums, 829-845
oral LD50, 1167, 1235
orchard mice, 8
Oriental rat fleas, 497
Oryzomys genus, 64

pack rats, 65, 341
parasites, 56, 314-330, 395-505, 636, 858
Pasteurella pestis, 488
Pasteurella tularensis, 859
Peludos, 709-729
Peromyscus leucopus, 422, 426, 519-541

Peromyscus species, 519–541.
  See also deer mice;
  white-footed mice
pet poisonings, 1155–1159
phosphine gas, 1232
pine voles, 6–7, 542–562
plague, 219, 321, 488–500,
  859
pneumonic plague, 498
pocket gophers, 10, 588–603
polybutene, 644
polycol foam, 789
popcorning, 271
porcupines, 604–617
prairie dogs, 618–624
Procyon. *See* raccoons
pygmy shrews, 924

Quintox, 1248–1252

rabbit plague, 859
rabbits, 846–875
  live trapping, 872–874
  protection of, 864
  repellents, 867–871, 875
  signs of, 861
rabies, 328
  in bats, 746–747, 749
  and raccoons, 915
  and skunks, 954
  and squirrels, 638
raccoons, 876–917
  activity periods, 882
  bait for, 910
  controlling, 893, 900–901,
    913
  invasion of structures, 878
  live trapping, 893–897, 903,
    908–909, 911–912, 914
  protection of, 890
  and rabies, 915
  relocating, 895
rat survey forms, 999
rat-bite fever, 307, 315,
  373–381
rats. *See also* Norway rats
  co-existence with house
    mice, 264
  cotton rats, 341, 506–518
  diastema, 60
  droppings, 58, 78
  Florida water rat, 63
  and leptospirosis, 363
  and multiple-catch traps,
    1109
  origins, 67
  resistance to anti-
    coagulants, 1265–1266,
    1272
  rice rats, 64
  roof rats, 200–240, 1222
  social grooming, 74
  teeth strength, 69

trapping, 1087–1088,
  1090, 1097
wood rats, 65, 341
rattlesnakes, 984–986
Rattus norvegicus. *See* Norway rats
Rattus rattus, 200–240, 1222
repellents, 1029, 1134–1137
  for bats, 784
  for chipmunks, 641, 645
  for moles, 805
  for rabbits, 867–871, 875
  for snakes, 996
  for squirrels, 641, 644–646
  resistance, 1263–1274
rice rats, 64
Rickettsia akari, 501
Rickettsia prowazeki, 481
rickettsialpox, 316, 501–505
rodent control, non-toxic,
  997–1143
  baiting, 1028, 1032,
    1277–1284
  best choices, 1000–1011
  black lights, 1140–1143
  building exteriors, 1026
  cars and dogs, 997–998
  elements of, 1034
  final rat or mouse, 1033
  glue boards, 659, 991,
    1076–1077, 1118–1133
  multiple-catch traps,
    1106–1117
  office buildings, 1025
  rat survey forms, 999
  repellents, 1029,
    1134–1137
  residences, 1024
  rodent odors, 1060–1065
  rodent stoppage,
    1035–1059
  signs of rodents,
    1012–1023, 1027
  sound devices, 1138–1139
  space needed for PCO,
    1031
  traps, 1066–1105
Rodent Gard, 1053
rodenticides, 1144–1276. *See
  also* bait
  active ingredient, 1194
  anticoagulant *versus* acute,
    1153
  anticoagulants, 668, 783,
    1164, 1253–1276
  Bitrex, 1169
  bromethalin, 1242–1247
  cholecalciferol, 1248–1252
  in food areas, 1154, 1192
  formulations, 120, 1193
  illegal, 1172–1174
  manufacturers, 1194
  minimum kill, 1171

mold inhibitors, 1188
pet poisonings, 1155–1159
secondary kill, 1152
success of, 1145
taste deterrents, 1169
terms defined, 1161–1168
tracking powder,
  1195–1216
trade names, 1194
zinc phosphide, 1217–1241
rodents, 1–79. *See also* non-
  commensal rodents; non-
  rodent pests
diastema, 71–72
dormancy, and blood sugar,
  50
droppings, 57–59, 77
gnawing and chewing, 61,
  70, 72
grooming, 73
hibernation, 49
parasites, 56
physical characteristics,
  45–48, 66, 70, 227
species of, 173
sylvatic rodents, 4
terms defined, 1–11
transportation modes, 55
trapping, 1066–1105
urine smells, 62, 76
roof rats, 200–240, 1222
RoPel, 644, 1134–1136
round-tailed muskrat, 63
roundworms, 916

salamanders, 595
Salmonella bacteria, 270, 382
Salmonella enteritidis, 383
Salmonella typhimurium, 383
salmonellosis, 326, 382–388
Sculopus aquatus, 810
secondary kill, 1152, 1163,
  1221
septicemic plague, 498
Sevin, 465
sewer rats. *See* Norway rats
ship rats, 200–240, 1222
shrews, 54, 918–943
Sigmodon, 341, 506–518
Sin Nombre virus, 331, 356
skunks, 944–971
Snake-A-Way, 996
snakes, 972–996
snap traps, 675, 679,
  1066–1067, 1076–1077,
  1081–1084, 1091,
  1104–1105
snowshoe rabbits, 857
sodium monofluoroacetate
  (1080), 1172–1173
Soricidae, 54, 918–943
sound devices, 1138–1139
Squirrel Away, 646

214 INDEX

squirrels, 625-659
star-nosed moles, 811-813
Streptobacillus moniliformis, 373
striped gophers, 596
Stuf-Fit, 1052
Suspend, 465
Sylvatic plague, 493-494
sylvatic rodents, 4
Sylvilagus. See rabbits
Sylvilagus floridanus, 846

Talpidae, 53, 795-828
tapeworms, 159, 319, 389-391
Target Animal Scent, 914
tartar emetic, 1234
taste deterrents, 1169
1080 (sodium monofluoroacetate), 1172-1173
1081 (fluoroacetamide), 1174
tetanus, 381
thallium sulfate, 1172
tick-removal tweezers, 430
ticks. See also deer ticks
  bites from, 407, 421
  black-legged, 398
  and cold winters, 436
  controlling with biological agents, 437
  dog ticks, 474
  and human granulocytic ehrlichiosis, 474
  lone star ticks, 399
  and Lyme disease, 398, 400-401, 407, 415-421, 424-425, 430, 432, 436, 463-466
  nymphal stage of, 417-419
  removing, 430
  wood ticks, 425

Tin Cat traps, 1117
tracking powder, 1168, 1195-1216
trade rats, 65, 341
training references, 1285-1293
trapping, 897, 899, 902, 904-907, 1066-1105.
  See also live trapping
  armadillos, 727
  bats, 786
  chipmunks, 675, 677, 679
  cleaning traps, 1073
  Conibear traps, 1099-1101
  examining traps, 1092
  in food areas, 1093
  glue boards, 659, 991, 1076-1077, 1118-1133
  mice, 284-285, 289, 294, 310, 358, 1068, 1075, 1080, 1090, 1096, 1106-1117
  moles, 815-828
  multiple-catch, 285, 1085-1086, 1106-1117
  opossums, 845
  pocket gophers, 602
  porcupines, 615-616
  raccoons, 897, 899, 902, 904-907
  rats, 107, 157, 512-513, 1087-1088, 1090, 1094, 1097
  securing traps, 1087
  shrews, 932
  skunks, 964-967, 969
  snakes, 994
  snap traps, 675, 679, 1066-1067, 1076-1077, 1081-1084, 1091, 1104-1105
  squirrels, 653

Tin Cat traps, 1117
  trigger traps, 1069, 1072
  voles, 560
  woodchucks, 702-703
Trichinella spiralis, 392
trichinosis, 323, 392-394
trigger traps, 1069, 1072
tularemia, 307, 859
typhus, 317, 481-487

urban plague, 219, 321, 488-500, 859

vampire bats, 758-759, 783
Vengeance, 1242-1247
vibrissae, 183, 1131-1132
voles, 6-7, 542-562

warfarin, 1260-1261, 1264-1265, 1272
water moccasins, 975
water rats. See Norway rats
Weil's Disease, 318, 359-366
wharf rats. See Norway rats
whistle pigs, 684-708
white-footed mice, 422, 426, 519-541
wood rats, 65, 341
wood ticks, 425
woodchucks, 684-708

Xenopsylla cheopis, 497

yellow-bellied marmots, 686
Yersinia pestis, 488

zinc phosphide, 109, 518, 1217-1241

Other Publications
by
Wildlife Control Consultant, LLC
Dealer Inquiries Welcomed

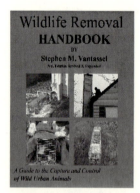

Wildlife Removal Handbook, 3rd ed.

205 pages of information on how to start and run your wildlife control operator business. Helpful for those who just want to learn how to control raccoons, skunks, squirrels, woodchucks and more like a professional.

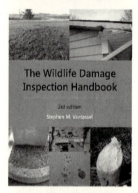

The Wildlife Damage Inspection Handbook, 3rd ed.

180 letter-sized pages. Color

This text teaches the how to read and interpret vertebrate animal sign and damage to structures and lawns and gardens. Don't guess what caused the problem, know what species caused the problem.

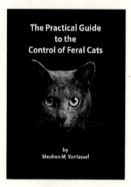

The Practical Guide to the Control of Feral Cats

103 pages of effective control information. Free-ranging cats are a growing environmental and public health threat around the country. Learn how to do your part to resolve human-feline conflicts.

Wildlife Control Consultant, LLC
WildlifeControlConsultant.com
©2019 Wildlife Control Consultant, LLC.

Made in United States
North Haven, CT
03 January 2022